INTEGRATED USE OF SPACE, GEOPHYSICAL AND HYPERSPECTRAL TECHNOLOGIES INTENDED FOR MONITORING WATER LEAKAGES IN WATER SUPPLY NETWORKS

Edited by **Diofantos G. Hadjimitsis, Athos Agapiou** and **Kyriacos Themistocleous**

Integrated Use of Space, Geophysical and Hyperspectral Technologies Intended for Monitoring Water Leakages in Water Supply Networks
http://dx.doi.org/10.5772/58836
Edited by Diofantos G. Hadjimitsis, Athos Agapiou and Kyriacos Themistocleous

Contributors

Diofantos Hadjimitsis

Published by InTech

Janeza Trdine 9, 51000 Rijeka, Croatia

Publishing Process Manager Danijela Duric
Technical Editor InTech DTP team
Cover InTech Design team

First published October, 2014
Printed in Croatia

Additional hard copies can be obtained from orders@intechopen.com

Integrated Use of Space, Geophysical and Hyperspectral Technologies Intended for Monitoring Water Leakages in Water Supply Networks , Edited by Diofantos G. Hadjimitsis, Athos Agapiou and Kyriacos Themistocleous
p. cm.
ISBN 978-953-51-1729-2

For EU product safety concerns:
IN TECH d.o.o., Prolaz Marije Krucifikse Kozulić 3, 51000 Rijeka, Croatia,
info@intechopen.com or visit our website at intechopen.com.

Meet the editors

Diofantos G. Hadjimitsis is a Full Professor and the Chair of the Department of Civil Engineering and Geomatics of the Cyprus University of Technology. He is also the Head of the Remote Sensing & Geo-Environment Lab. He has more than 300 publications in journals, conference proceedings and chapters in books in the field of remote sensing, space applications, UAV, geo-information and GIS for environmental surveillance /environmental monitoring, geomatics and civil engineering.
He undertook his PhD and MPhil (1996-1999) in the field of Satellite Remote Sensing intended for water quality monitoring in water treatment reservoirs from the University of Surrey, Department of Civil Engineering (UK). His PHD has been sponsored from Thames Water Utilities (UK). During his doctoral studies he received an ORS Scholarship and a Faculty of Engineering Scholarship. He obtained a First Class BEng. (Honours) Degree in Civil Engineering from Surrey University (1994-1996). Diofantos was awarded the Felix Pulzer Memorial Prize (Surrey Univ.) and National Guard Prize (HTI). He obtained a First class HND Diploma in Civil Engineering from the Higher Technical Institute (1989-1992). Diofantos obtained a distinction MSc award in Real Estate and Property Management (RICS), from the University of Salford (UK) (2005-2007). Diofantos worked in the past at the University of Surrey (UK), University of Southampton (UK), and Frederick University and as a managing director in a consultancy company working in the fields of remote sensing and GIS. He is founder member and Vice-President of the Cyprus Remote Sensing Society. He is the Chairman of the 'RSCY International conferences -Remote Sensing and Geo-information of the Environment'. Diofantos is a Senior Member of SPIE (2014). He is the member of the National Experts appointed by the Cyprus Government for investigating the potential of Cyprus to join the European Space Agency (ESA). He is the member of the editorial team of the Central European Journal of Geosciences.

Dr Athos Agapiou is a senior researcher of the Department of Civil Engineering and Geomatics of the Cyprus University of Technology. He is also the Head of the Cultural Heritage and Archeology Division of the Remote Sensing & Geo-Environment Lab. He has more than 25 publications in journal papers and more than 80 publications in conference proceedings, books etc. He received his MEng from the National Technical University of Athens (NTUA). He continued his studies at the National Technical University of Athens (NTUA) where he obtained his MSc Degree in Geoiformatics. Athos also holds a MA in the field of Archaeology from the University of Cyprus.

He received his PhD from Cyprus University of Technology in the field of Remote Sensing. His scientific interests range in the fields of Remote Sensing; Geographical Information Systems; Remote Sensing Archaeology; Archive Aerial Image processing; Archaeolandscapes; Natural and Anthropogenic hazards; Ground Spectroscopy; Geometric Documentation of CH and sites etc. Athos he has been involved in more than 25 European and national funded projects since 2007 and he is also a reviewer in more than 15 journals related with Remote Sensing.

Dr. Kyriacos Themistocleous is a senior researcher of the Department of Civil Engineering and Geomatics of the Cyprus University of Technology. He is also the Head of the UAV Division of the Remote Sensing & Geo-Environment Lab. He has more than 150 publications in Cyprus University of Technology and he has been involved in more than 25 European and national funded projects since 2008. He has over 150 publications in the field of environmental sciences, geophysics, water management, UAV, remote sensing, spectroscopy, geographical information systems (GIS), and atmospheric pollution. He is the Head of the Environmental and Remote Sensing unit of the Remote Sensing and Geo-Environment laboratory at the Cyprus University of Technology. His research interests include environmental studies, geophysics, GIS, aerial and satellite remote sensing, spectroscopy, Lidar, atmospheric correction and cultural heritage. He is member of the Board of Directors of Cyprus Scientific and Technical Chamber and President of the Cyprus Remote Sensing Society.

Contents

Preface

Cyprus faces the growing problem of water shortages. At a time when huge efforts are being undertaken to secure sufficient drinking water to meet the needs of the island, the water supply network leakage problems are becoming increasingly important for Cyprus. On site observation is the most common way of monitoring pipe network in Cyprus. However, this procedure which includes data collection, periodical observations, and multivariate risk assessment analysis, is difficult to be accomplished with the traditional methods since it is time consuming, expensive and the monitoring is localized. Furthermore, part of the network is located at inaccessible areas, far away from the main road network and urban areas.

A complete geo-information system providing the exact location, characteristics and relevant data for the water mains does not exist, making the leakage monitoring procedure even more complicated. This book presents the outcomes of two research projects which both interrelated with the smart management of water utilities. Indeed, Chapter 1, 2,3 presents the outcomes from the 'Water leaks' project that aims to use a combination of new, innovative technologies for a complete system of water distribution network leakage detection and management, by combining Global Navigation Satellite Systems (GNSS), Geographical Information Systems (GIS), Satellite Remote Sensing techniques as we Geophysical surveys, Unmanned Aerial Vehicles (UAV) and spectroradiometric measurements. Finally, Chapter 4 presents the overall methodology and an overview of the developed infrastructure, named as 'HYDROGIS Lab' which is used for the modeling of the water supply network by integrating ground penetrating radar (GPR), GIS and GPS.

Diofantos G. Hadjimitsis
Full Professor and the Chair of the Department of Civil Engineering and Geomatics,
Cyprus University of Technology, Cyprus

Dr Athos Agapiou
Department of Civil Engineering and Geomatics
Cyprus University of Technology, Cyprus

Dr. Kyriacos Themistocleous
Department of Civil Engineering and Geomatics
Cyprus University of Technology, Cyprus

Chapter 1

Introduction

Skevi Perdikou, Kyriacos Themistocleous,
Athos Agapiou and Diofantos G. Hadjimitsis

Additional information is available at the end of the chapter

http://dx.doi.org/10.5772/59521

1. Introduction — The problem of water leakages

Water is a daily necessary resource for life, health, economic development and the ecosystem all over the world. As water is precious to everyone, its availability and quality are essential. Climate change, droughts, water shortages and population growth are increasing the strain on existing water resources, thereby increasing the necessity to preserve and avoid water wastage through effective management and reduction of water losses.

A significant amount of water is lost in the water supply system. Water leakages have been a major problem for many regions around the world (Weifeng et al, 2011). In some countries water loss due to water leakages in the supply network exceeds 40% of the water in the supply system (Fig 1). Reduction of water leakages is an important goal for many countries in the world (Vairavamoorthy and Lumbers, 1998), as it will mean a reduction in the amount of money and energy required on producing and pumping water and also satisfaction of consumer needs through improved reliability of the system (Ayala–Cabrera et al, 2013).

The Eastern Mediterranean countries are expected to be affected adversely by climate change. Similar to other countries in this region, Cyprus has semi-arid climate with limited water resources (Zachariades, 2010). This, combined with the high water demand on the island, resulted in Cyprus having the highest Water Exploitation Index across Europe WEI (Fig 2). A WEI above 20 % implies that a water resource is under stress and values above 40 % indicate severe water stress and clearly unsustainable use of the water resource (Raskin et al., 1997).

Traditional field survey methods are costly and time-consuming, as the timely identification and monitoring of leakages in the water supply network is a very complex task (Huang et al. 2010). Over the years, different non-intrusive techniques have been used to monitor and identify water leakage. These can be grouped as (a) visual techniques such as closed-circuit television, (b) electromagnetic and radio frequency techniques e.g. magnetic flux leakage,

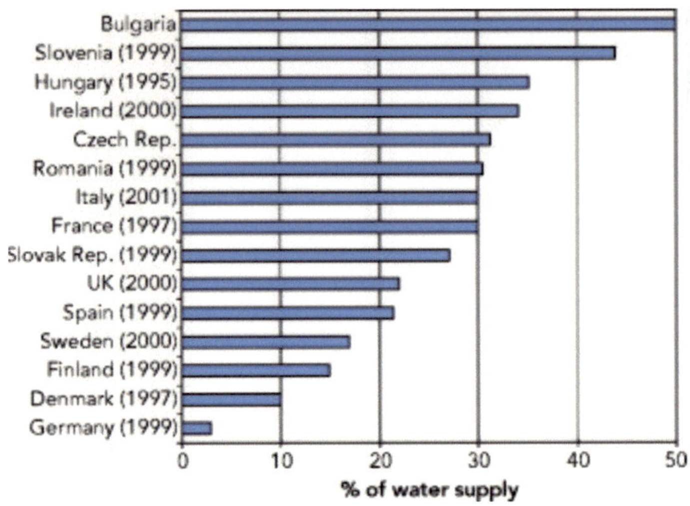

Figure 1. Losses from urban water networks. (): Year of data. (http://www.eea.europa.eu/legal/copyright). Copyright holder: European Environment Agency (EEA, 2008).

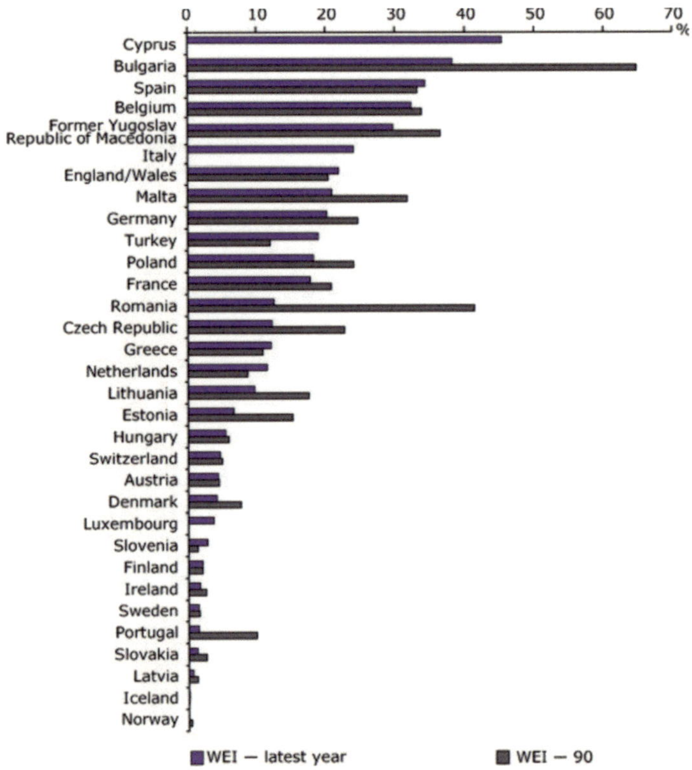

Figure 2. Total water abstraction per year as a percentage of the total long-term renewable resource. (http://www.eea.europa.eu/legal/copyright). Copyright holder: European Environment Agency (EEA, 2008).

ground penetrating radar, (c) acoustic and vibration techniques like sonar, vibro-acoustics and (d) others such as infrared thermography, laser surveys (Ayala-Cabrera et al. 2013, Hadjimitsis et al. 2009, Eyuboglu et al. 2003).

2. Remote sensing and radar technology for water leakage detection

Remote sensing has been used for water management purposes over the years. Recent remote sensing techniques have been used in water leakage detection studies mainly through the use of Vegetation Indices (VI). Agapiou et al. (2012a) has divided VI into five main categories according to equation or the use of each index, which include broadband indices, narrowband indices (hyperspectral), leaf pigment indices, stress indices and water stress indices. They stated that VI can be simply divided according to the wavelength characteristics used in their formula (broadband and narrowband indices). A study by Pickerill and Malthus (1998) showed that the use of different VI and single bands were required for the identification of the water leaks. Work by Huang et al. (2009) indicated that the use of red, NIR and thermal bands is highly consistent with the field observations in the detection of water leakage in irrigation distribution networks. For their study, they used airborne multispectral remote sensing imagery with high-resolution imaging sensors in the visible, NIR and thermal infrared wavelengths.

This book describes the combination of satellite imagery, in-situ spectroradiometric data and radar techniques for the identification of water leakages in the water supply network in rural areas in Cyprus. The water mains network covers great distances in rural and sometimes inaccessible areas, and for most of the cases in Cyprus, the exact position of the network is not known. Therefore, tracing water leakages is a very costly, time consuming task adding to the already immense problem of water sustainability on the island. A complete geoinformation system providing the exact location, characteristics and relevant data for the water mains does not exist, making the leakage monitoring procedure even more complicated. This book presents a holistic approach combining new technologies for a complete system of water distribution network leakage detection management, which can be used in different countries to effectively identify and monitor water leakages.

3. Study-area

In order to identify and monitor water leakages, 3 case studies are described in this section: (a) Frenaros-Choirokoitia (b) Lakatamia and (c) Germasogeia water pipe network. The first case study Frenaros-Choirokoitia (Fig 3) is a major rural pipeline in Cyprus, which runs from the *Choirokoitia* area to the *Frenaros* area. The existing pipeline, which passes through the central and central-east part of Cyprus, has a length of over 65 km. The pipeline is located 1-3 meters below ground surface. Various geological formations, generating several soil types including calcaric cambisols, calcaric regosols, and epipetric calcisols, exist in the area, while the higher and lower elevation of the pipeline (ground surface) is estimated to vary between 10 m and 200 m above sea level. From 2007 to 2010, three major leakages were observed along different sections of the pipe (Fig 4). The locations of these leakages were not detected at once after the leakage occurred. The leakages occurred during 2007, 2008 and 2010; further details for these events are described later in the book.

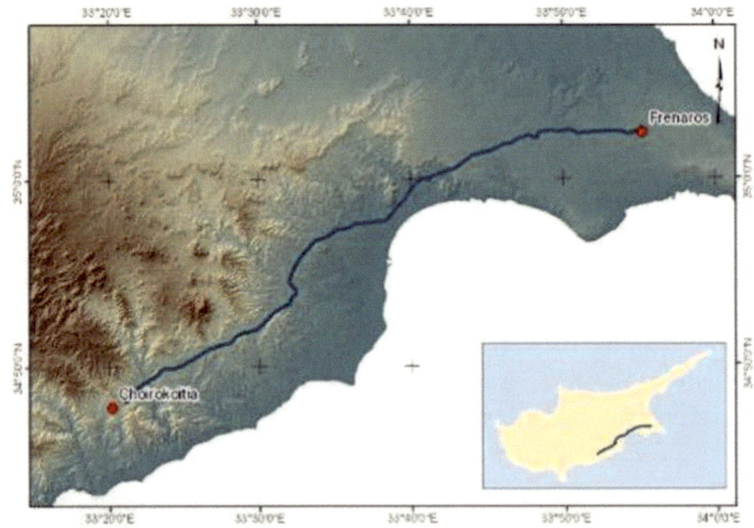

Figure 3. The "Choirokoitia-Frenaros" water pipe

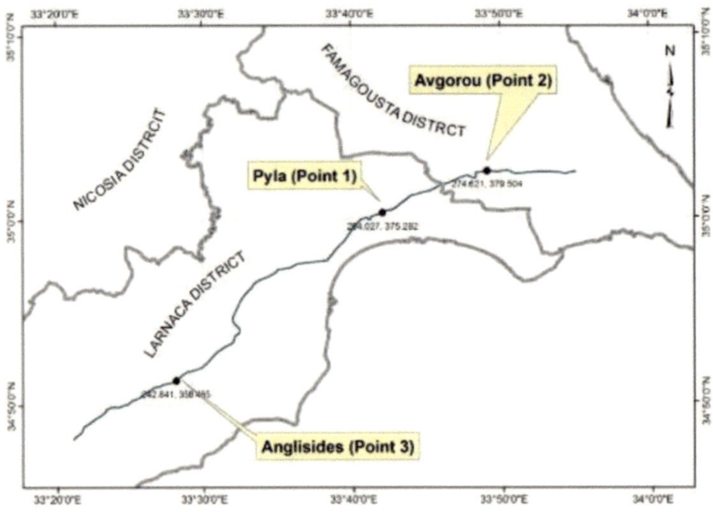

Figure 4. Points 1-3 indicate the spots were water leakages have been reported in the *"Choirokoitia-Frenaros"* water pipe.

3.1. Lakatamia water pipe

The Lakatamia water pipe (Fig 5) has a length of less than 5 km and has been systematically reported as problematic due to several leakages and it is therefore no longer in use by the local authorities. The pipe runs through urban and rural areas. A section of the pipeline, with a length of over 2km and located in a rural area, has been used for the detection of leakages. Since the existing water pipe is not currently in use, it was necessary to fill the pipe periodically with water to observe the effectiveness of such remote sensing non-contact techniques. The water pipe is made of UPVC and has 315mm diameter. It is between 1.80m and 2.00m below the ground surface and runs along the *Pediaos* River for a large part of its length. It is currently not being used due to water leakages occurring throughout almost the entire length of the pipeline. Information regarding the specific dates of events is not available from local authorities.

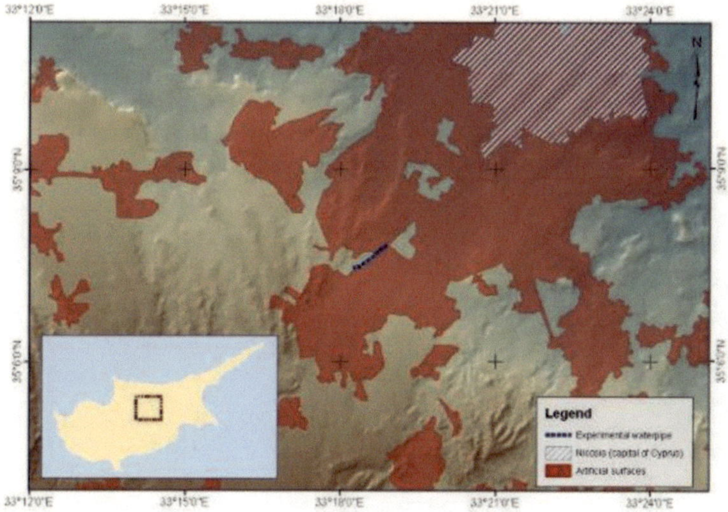

Figure 5. The *"Lakatamia"* water pipe

3.2. Germasoyeia-Moutagiakia pipeline

The third case study was the "Germasoyeia-Moutagiakia" pipeline in the southern part of Limassol District in Cyprus. This pipeline runs through rough terrain, making visual inspections difficult. This pipeline was selected in order to evaluate the results from the satellite images.

4. Methodology

For all three case studies, the first step was to georeference the water pipes under examination. For the "Chroirokoitia-Frenaros" water pipe, all data regarding the characteristics of the pipeline were collected in CAD format from the Water Development Department-which is the responsible authority for the water pipe. In addition, the coordinates of the leakages were collected and added in a GIS environment. The GIS environment allows the researchers to identify potential areas which are suitable for the detection of the event, so they can be identified on satellite images. Soil, land use and meteorological data were also included in the GIS environment. A similar approach was followed for the "Germasoyeia-Moutagiakia" pipe line. The CAD file was converted into a :shp file and imported into a GIS database. For the pipeline in the area of Lakatameia, GPS measurements were carried out to map the path of the pipeline (Figure 6; Figure 7). Engineers from the municipality of Lakatameia also indicated problematic spots over the buried pipeline, which were recorded in the data. Figure 7 shows the pipeline as mapped from the GPS measurements (with red line, background Google Earth).

Figure 6. GPS measurements in the Lakatameia pipeline.

Figure 7. Pipeline as mapped from the GPS measurements (with red line, background Google Earth).

5. Resources

The resources are grouped into three main categories: (a) high resolution satellite data used for the *"Southern Conveyor Project"* area; (b) spectroradiometric ground data used for the *"Lakatameia"* pipeline and (c) medium resolution satellite data used for the *"Choirokoitia-Frenaros"* pipeline. In addition, spectroradiometric hyperspectral measurements were carried out using the GER 1500 field spectroradiometer (Figure 8a). The GER 1500 spectroradiometer records electromagnetic radiation between 350 nm to 1050 nm (visible and near infrared part of the spectrum) A calibrated Spectralon panel, with ≈100% reflectance, was also used simultaneously to measure the incoming solar radiation. The Spectralon panel measurement was used as a reference, while the the crops were used for spectroradiometric measurements. Therefore, reflectance for each measurement can be calculated using the following equation (1):

$$\text{Reflectance} = \left(\text{Target Radiance} / \text{Panel Radiance}\right) \text{ x Calibration of the panel} \tag{1}$$

In order to avoid any errors due to changes in the prevailing atmospheric conditions (Milton et al. 2009), the measurements over the panel and the target were taken within minutes of each other. The coordinates of the measurements were mapped using a Global Navigation Satellite Systems (GNSS) (Figure 8b).

Figure 8. a) (left): The GER 1500 spectroradiometer used for the collection of ground measurements and b) (right): the GNSS used for mapping the pipeline

In addition, spectroradiometric measurements were taken from a low altitude system (Figure 9). The spectroradiometer was attached to the air balloon and raised over the pilot study area. Measurements were taken at several heights in the pilot study area and also in the surrounding area in order to compare their spectral signature profiles. As the airborne system was raised, the pixel size of the ground increased.

Spectroradiometric data have also been used in order to form thresholds of reflectance and vegetation indices (VI) that indicate a 'water leakage suspicious' event. To do this, a great number of measurements were recorded and different vegetation indices were determined for soil and vegetation under 'dry' and under 'wet' conditions. This allowed the formation of 'thresholds' which imply a 'water leakage suspicious' event when reflectance values and vegetation indices fall within these thresholds. By observing satellite data and specifically reflectance and vegetation indices values, the authorities will identify 'suspicious' areas for water leakage. This will minimise the in-situ visits to only these suspicious areas, significantly reducing cost and time for the water authorities and gaining reliability and satisfaction for the consumers.

The final outcome of the various studies will be a geographical information system (GIS) that produces maps indicating specific locations where a water leakage may have occurred. As stated by Chorley (1987), a GIS is a system for collecting, storing, managing, analyzing and presenting data related to space. The core of a complete GIS system consists of three basic components that are in constant balance and interdependence. These are machines (hardware), algorithms (software) and available data (resources).The possibilities offered by GIS make

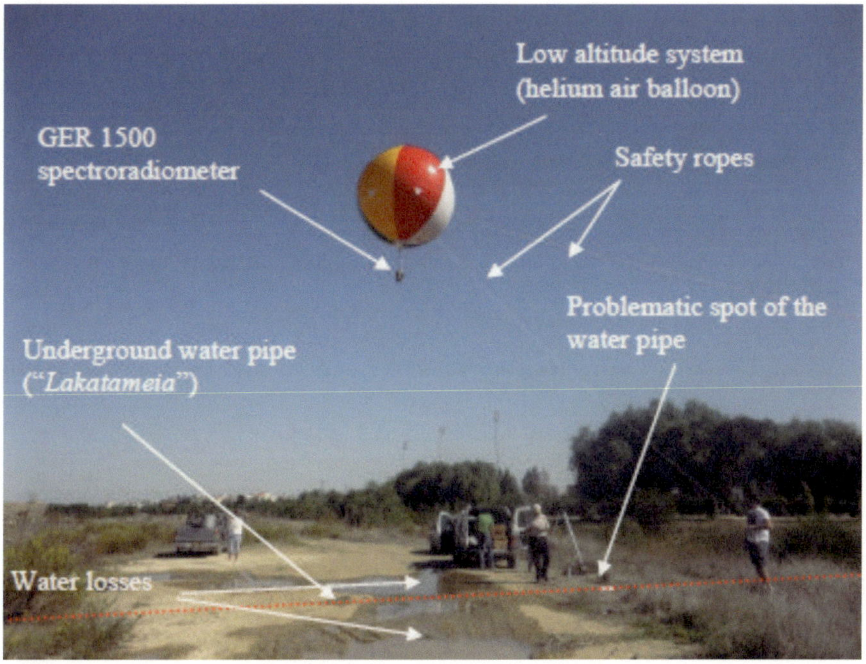

Figure 9. The low altitude system for spectroradiometric data

them a valuable tool, as the ability of GIS to able to correlate different information in a spatial context, thereby creating relationships and conclusions. GIS can be used to emphasize the spatial relationships among the objects being mapped. Therefore, GIS can reveal important new information that leads to better understanding and decisions.

In order to design a GIS database, some basic steps must be followed. These steps include: (1) exterior design of the model, describing the real world (depending on the needs of users), (2) conceptual design, where a generalization of the real world is performed and final 3) the logical design where the implementation is made at a computer.

For the data of the three case studies described in this book, a customized GIS was developed to monitor changes regarding the pipelines both in space and in time. The continuous updating and monitoring of water pipelines can assist the public authorities to examine changes over time, based on a continuously feedback of the GIS database. Temporal GIS is an emerging capability for integrating temporal data with location and attribute data, enabling temporal visualization and ultimately temporal analysis.

Acknowledgements

The results reported here are based on findings of the Cyprus Research Promotion Foundation project "ΑΕΙΦΟΡΙΑ/ΦΥΣΗ/0311(ΒΙΕ)/21": Integrated use of space, geophysical and hyperspectral technologies intended for monitoring water leakages in water supply networks in Cyprus. The project is co-funded by the Republic of Cyprus and the European Regional Development Funds of the EU.. Thanks are also given to the Remote Sensing and Geo-Environment Laboratory of the Department of Civil Engineering & Geomatics at the Cyprus University of Technology for its continuous support (http://www.cut.ac.cy).

Author details

Skevi Perdikou[1*], Kyriacos Themistocleous[2], Athos Agapiou[2] and Diofantos G. Hadjimitsis[2]

*Address all correspondence to: s.perdikou@frederick.ac.cy

1 Department of Civil Engineering, Faculty of Engineering, Frederick University, Lefkosa, Cyprus

2 Department of Civil Engineering and Geomatics, Faculty of Engineering and Technology, Cyprus University of Technology, Cyprus

References

[1] Agapiou, A., Hadjimitsis, D.G., Alexakis, D. and Sarris, A.. Observatory validation of Neolithi tells ("Magoules") in the Thessalian plain, central Greece, using hyperspectral spectroradiometric data. *Journal of Archaeological Science,* 2012a. 39(5): pp. 1499-1512. doi: 10.1016/j.jas.2012.01.001

[2] Ayala-Cabrera, D., Herrara, M. Izquierdo, J., Ocana-Levario, S.J. and Perez-Garcia, R.. GPR-based water leak models in water distribution systems. *Sensors,* 2013. 13: pp. 15912-15936. doi: 10.3390/s131215912

[3] Department of the Environment. Handling Geographic Information. *Report of the Committee of Enquiry Chaired by Lord Chorley,* 1987. London: HMSO

[4] European Environment Agency. 2008. Available from: http://www.eea.europa.eu/legal/copyright.

[5] Eyuboglu, S., Mahdi, H., and Al-Shukri, H.. Detection of Water Leaks using Ground Penetrating Radar, in:*Proceedings of the 3rd International Conference on Applied Geophysics.* December 8, 2012. Orlando, Florida. Orlando Floria: Environmental and Engineering Geophysical Society. 2003.

[6] Hadjimitsis, D.G., Themistocleous, K., and Achilleos, C.. Integrated use of GIS, GPS and Sensor Technology for managing water losses in the water distribution network

of the Paphos Municipality in Cyprus, in:*STATGIS*. 17-19 June, 2009. Milos, Greece. Milos, Greece: 2009.

[7] Huang, Y., Fipps, G., Maas, S.J. and Fletcher, R.S. Airborne remote sensing for detection of irrigation canal leakage. *Irrigation and Drainage*, 2009. doi: 10.1002/ird

[8] Milton, E.J., Schaepman, M.E., Anderson, K., Kneubohler, M and Fox, N.. Progress in Field Spectroscopy. *Remote Sensing of Environment*, 2009. 113: pp. 92-109.

[9] Pickerille, J.M., and Malthus, T.J.. Leak detection from rural aqueducts using airborne remote sensing techniques.. *International Journal of Remote Sensing*, 1998. doi: 1924272433

[10] Raskin, P., Gleick, P.H., Kirshen, P., Pontius, R.G and Strzepek, K.. *Comprehensive assessment of the freshwater resources of the world-Document prepared for the 5th session of the United Nations Commission on Sustainable Development*. Stockholm Environmental Institute, Sweden. 1997.

[11] Vairavamoorthy, K. and Lumbers, J.. Leakage reduction in water distribution systems: optimal valve control. *Journal of Hydraulic Engineering*, 1998. 124(11): pp. 1146-1154.

[12] Weifeng, L., Wencui, L., Souxiang, L., Jing, Z., Ruiping, L., Qiuwen, C., Zhimin, Q. and Jiuhui, Q. Development of systems for detection, early warning and control of pipeline leakage in drinking water distribution: A case study. *Journal of Environmental Sciences*, 2011. doi: 2318161822

[13] Zachariades,. Residential water scarcity in Cyprus: Impact of climate change and policy options. *Water*, 2010. 2doi: 10.3390/w2040788

Chapter 2

Investigation of Ground Remote Sensing Techniques for Supporting an Early Warning Water-Leakage System

Athos Agapiou, Dimitrios D. Alexakis,
Kyriacos Themistocleous, Apostolos Sarris,
Skevi Perdikou, Chris Clayton and
Diofantos G. Hadjimitsis

Additional information is available at the end of the chapter

http://dx.doi.org/10.5772/59531

1. Introduction

The decrease of water availability is a major global problem that is increasing in intensity. Since water resources have been substantially reduced over the past years, control over water distribution is now considered imperative. For this reason, water supply networks are more closely monitored and significant efforts are made to reduce the effects of leakages. In addition to the design and maintenance of water distribution systems, researchers are also focused on the improvement of early detection and rapid response of a leakage. Research indicates that water supply network may lose up to 20-30% of water, with the main cause being water leaks (Cheong, LC 1991). In some networks, the loss can reach more than 50% (AWWA, 1987).

Systematic leakages can lead to significant losses in both water as well as financial resources. Water pipe networks, regardless of age, often present problems of water leakage, resulting in large losses of precious drinking water. Therefore, there is an urgent need to design and build systems that can detect the presence and location of leaks in water pipes networks. The detection of hidden leaks in underground water supply networks requires the use of instruments designed specifically for this purpose by a trained operator.

The most widely applied technique for the detection of hidden leaks is the acoustic method. Additional methods for leak detection also include thermography, remote sensing, geophysics etc. Leaks from pipelines under pressure create a whistling characteristic noise, which is transmitted by the water itself. Hunaidi and Chu (1999) used the acoustic method for leakage detection, which focused on various types of leaks under controlled conditions. Hunaidi and

Giamou (1998) conducted a survey on the effectiveness of acoustic methods and the dynamics of alternative non-acoustic methods for leak detection in plastic pipes, using penetrating ground radar to assess the potential leakage. The survey was conducted in a specially constructed facility where different types of leaks could be simulated under controlled conditions. The facility had a 200m long tube made from PVC, 150mm diameter, which was buried in soft muddy soil to a depth 2.4m. The research was performed using a radar antenna with frequencies of 50, 100, 200, and 450 MHZ.

Pickerill and Malthus (1998) study of water leakage detection using airborne remote sensing data used the analysis of soil moisture and vegetation biomass based on thematic maps. The researchers were able to detect two leaks from the surrounding environment of the Aqueduct of the North West England. Huang and Fipps (2002) used an airborne thermal sensor, including GPS receiver, for leakage detection over irrigation canals and pipelines. The sensor was able to record surface temperature as digital values ranging from 0-255 (8 bit). Based on the data, 45 areas where identified as potential leakage sites. Each site was mapped using Geographical Information System of the area. From the 45 areas, 11 sites were examined in-situ, with a success rate of 91%. The other sites were specific channels with cracks that were likely sources of leaks. The study concluded that the thermal imaging method was a very promising technology for the evaluation of irrigation channels and leakage as it is a rapid and cost effective method of leak detection for irrigation channels. In a similar study, Huang and Fipps (2002) conducted a study in the Rio Grande Valley of Texas using airborne multispectral remote sensing for leak detection and possible identification of an irrigation network. A multispectral imaging system combining red, NIR and thermal sensors was used to collect image data in over 24 selected. During the flight, the three cameras (red, NIR and thermal) were synchronized to obtain the same area. Image processing included both image fusion as well as image georegistration. The combination of red, NIR and thermal sensors were effective for the determination of leakage areas. The study found that airborne multispectral imaging can be used for assessing both the conditions of the channel as well for the detection of leaks in irrigation distribution networks.

Shih and Jordan (1993) examined soil moisture detection using Landsat satellite images. The methodology involved daily temperature data and soil moisture measurements recorded with ground-based observations. The results indicated that the percentage content of soil moisture was inversely proportional to the temperature at the soil surface. The thermal infrared (IR) data from Landsat band 6 was classified into four land use classes using GIS data. The four main categories were agriculture, irrigation, urban and forest-wetlands. The thermal data were also used to evaluate four soil moisture conditions (water, very wet, wet, dry) and were used in each of the four classes mentioned above. The integration of Landsat images using GIS can be used to assess regional soil moisture conditions which can be adapted to identify water leak detection. Naimullah (2009) used satellite images for pipeline leak detection, further to the classification of land use in Mempatih, Pahang, Malaysia. The objective of the study was the evaluation of vegetation indices such as NDVI images using SPOT-5 satellite data. The final results showed a correlation of 81.02% for NDVI to identify the leakage problem.

The research presented above, along with the studies discussed in this book, highlight the significant contribution of satellite imagery for the detection of leakages. In this chapter, further investigation of the use of ground remote sensing to identify water leakage in Cyprus is presented.

2. Methodology and resources

In the study of leakage detection in Cyprus, ground spectroradiometers and a digital infrared thermometer were the main tools used to monitor and investigate potential leakage events. In order to use ground spectroscopy for the detection of leakage problems, several measurements were made in a controlled environment. The experiment was conducted at the Agricultural Research Institute, located in Acheleia, Cyprus. The area of interest measured approximately 11 x 7 meter, as indicated in Figure 1. A grid of 1 x 1 meter was created to take measurements. Measurements were taken with a ground spectroradiometer before and after the flooding of the pipeline in order to compare the difference observed in the spectral signatures profiles. Due to weather conditions, frequent calibration of the incoming solar radiation using the Spectralon panel was necessary. The Spectralon panel is a Lambertian surface with 99.99% reflectance. Areas with shadows were excluded in this study in order to avoid misleading results. The measurements were taken on clean and dry ground.

Figure 1. Area of interest. A pipeline was buried at a depth of approximately 1 meter.

In this area, a pipeline was buried 1 meter below surface. Cracks were produced in the pipeline in order to create an artificial leakage event. The end of the pipeline was connected to a valve. Figure 2 shows the area of interest approximately one hour after the artificial leakage event.

Figure 2. Leakage as observed from the surface

The GER 1500 PC Graphing Application Software was used for the analysis of the measurements and to plot the spectral signatures. For the calculation of the vegetation indices, a Matlab code was applied. The maps were created using the ArcGIS software. Based on the Relative Response Filters of the Landsat 7 ETM+sensor, the narrow band reflectance recorded from the spectroradiometer was recalculated into the above satellite sensor. By using the broadband reflectance values (Bands 1 – 4; Blue / Green / Red / Near Infrared) several vegetation indices were calculated as shown in table 1.

no	Vegetation Index	Equation
1	EVI (Enhanced Vegetation Index)	$2.5\ (p_{NIR} - p_{red})/(p_{NIR} + 6\ p_{red} - 7.5\ p_{blue} + 1)$
2	Green NDVI (Green Normalized Difference Vegetation Index)	$(p_{NIR} - p_{green})/(\ p_{NIR} + p_{green})$
3	NDVI (Normalized Difference Vegetation Index)	$p_{NIR} - p_{red})/(p_{NIR} + p_{red})$
4	SR (Simple Ratio)	p_{NIR}/p_{red}
5	MSR (Modified Simple Ratio)	$p_{red}/(p_{NIR}/p_{red} + 1)$^0.5

no	Vegetation Index	Equation
6	RDVI (Renormalized Difference Vegetation Index)	$(p_{NIR} - p_{red})/(p_{NIR} + p_{red})$^1/2
7	IRG (Red Green Ratio Index)	$p_{Red} - p_{green}$
8	RVI (Ratio Vegetation Index)	p_{red}/p_{NIR}
9	MSAVI (Modified Soil Adjusted Vegetation Index)	[2 p_{NIR}+1-[(2 p_{NIR}+1)^2 -8(p_{NIR} - p_{red})]^1/2]/ 2
10	ARVI (Atmospherically Resistant Vegetation Index)	$(p_{NIR} - p_{rb})/(p_{NIR} + p_{rb})$, $p_{rb} = p_{red} - \gamma\,(p_{blue} - p_{red})$
11	GEMI (Global Environment Monitoring Index)	n(1-0.25n)(p_{red} -0.125)/(1 − p_{red}) n = [2(p_{NIR}^2- p_{red}^2)+1.5 p_{NIR}+0.5 p_{red}]/(p_{NIR}+ p_{red} +0.5)
12	SARVI (Soil and Atmospherically Resistant Vegetation Index)	(1+0.5) (p_{NIR} - p_{rb})/(p_{NIR} + p_{rb} +0.5) $p_{rb} = p_{red} - \gamma\,(p_{blue} - p_{red})$
13	OSAVI (Optimized Soil Adjusted Vegetation Index)	$(p_{NIR} - p_{red})/(p_{NIR} + p_{red} +0.16)$
14	DVI (Difference Vegetation Index)	$p_{NIR} - p_{red}$
15	SR × NDVI (Simple Ratio x Normalized Difference Vegetation Index)	(p_{NIR}^2– p_{red})/(p_{NIR} + p_{red}^2)

Table 1. Vegetation indices used in this study.

A handheld digital infrared thermometer was also used to measure the ground temperature of the area. By using the laser sighting, measurements were taken at 1 meter intervals, before the leakage, during the leakage as well after the leakage event. Although the weather may have varied before, during and after the leakage, this had no impact on the results since the maps presented and discussed are based on relative differences in temperature. After the recording of the measurements, the ArcGIS was used to plot the data. Several interpolation methods were used, including the inverse distance to the power (IDW).

3. The use of UAV for detecting water leakages

Unmanned aerial vehicles (UAVs) have been used for monitoring purposes of leakages events in several semi-arid areas of Cyprus. Due to the low cost and high quality image provided, UAVs are increasingly being used for field work. Research indicates that aerial remote sensing and imaging can be conducted using large scale low-altitude imagining and geospatial information (Colomina and Molina, 2014; Cho et al, 2013; Mayr, 2013; Petrie, 2013). In this study, a UAV platform was used to take aerial images of the study site. The UAV was equipped with RGB and VNIR cameras in order to document and detect water leakages from a low-altitude position. In this study, an octo-copter with 8 motorized propellers, including a GPS, piezo gyros, compass, altitude control, telemetry, acceleration and barometric sensor was used (figure 3.) The copter was used to take photographs of the water leakage events. Figure 4 features an aerial view of the Mandria water leakage site, while Figure 5 clearly shows the Lakatameia underground pipeline via aerial photography, with the continuation of the pipeline in yellow.

Figure 3. Oktocopter UAV

Figure 4. Leakage event in Mandria as observed from UAV

http://dx.doi.org/10.5772/59531

Figure 5. Lakatameia underground pipeline as observed from UAV

It was found that, in some cases, the UAV can identify water leakages and water pipelines in different ways such as increased vegetation along the pipeline where the water leakage occurs, change of soil characteristics along the pipeline route (Figure 5), as well as changes in the reflectance value of soils in the water leakage area. UAVs can also be used with thermal and infrared cameras to identify changes in temperature due to water leakage. An additional benefit of UAV platforms is that they can be used to survey large areas to provide overall coverage of the area or it can be used for areas that are inaccessible. The integration of UAV with other techniques is a useful method to cross-validate the leakage event.

4. Results

4.1. Digital Infrared Thermometer results

Figure 6 indicates the results from the ground temperature measurements before the leakage event. The results show that the minimum temperature values are recorded around the pipeline. This is most likely due to the presence of soil moisture resulting from a prior leakage, which occurred a month prior to the measurements. Similar results were also observed in the top of Figure 6, since the olive trees present in the area tend to hold moisture in the soil. In comparing the soil temperature over the pipeline against surrounding area, a relative difference of 2 °C was observed. This difference is logical, since from the last leakage event the soil temperature tends to be close to its original (background) value.

Figure 6. Ground temperature over the area of interest as recorded from the digital laser thermometer, before the leakage event. Rectangle indicates the location of the buried pipeline. Blue values indicates temperatures 14-16°C while the cyan colour temperatures 16-18°C.

Figure 7 demonstrates the results during the leakage event. The lowest surface temperature is shown in the western part area over the pipeline. In this section, the leak was initially observed during the leakage event. The soil temperature for the rest of the area was relatively high. These values indicate that moisture was not detected. The soil temperature also indicated a relative difference of about 6 °C of the leakage event and the rest dry area. Figure 8 features a photograph taken during the leakage event, as well the results from this area. The temperature measurements were able to map the problematic area with high accuracy and thereby define the leakage event. After the leakage event, the soil temperature decreased in the area over the buried pipeline. As shown in Figure 9, the entire pipeline had a tendency to present lower temperature values, including the leakage zone. The difference recorded in this step was estimate to be 8°C.

In comparing the above figures, it appears that the soil temperature over the buried pipeline provided lower values before, during and after the leakage event. The relative temperature difference in this zone in relation to rest of the study area was initially small, then subsequently decreased and finally further decreased. The relative difference temperature of this zone prior to and after the leakage indicates a 60% reduction in relation to the surrounding area.

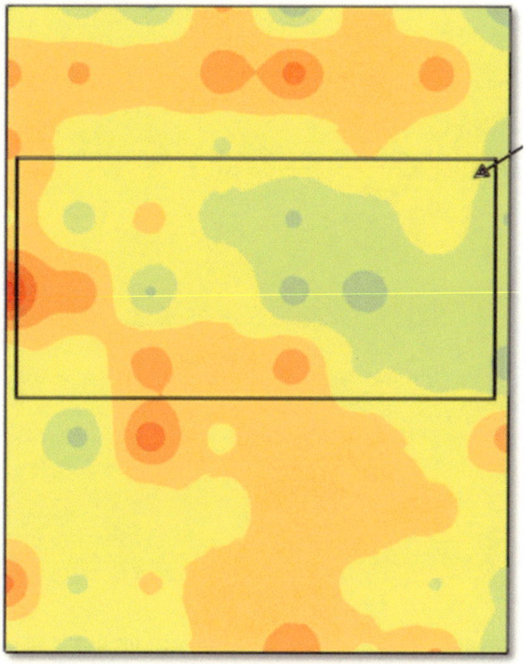

Figure 7. Ground temperature over the area of interest as recorded from the digital laser thermometer, during the leakage event. Rectangle indicates the location of the buried pipeline. Blue values indicates temperatures 14-16°C, 24-26°C are shown in yellow while orange colour highlight values from 34-36°C

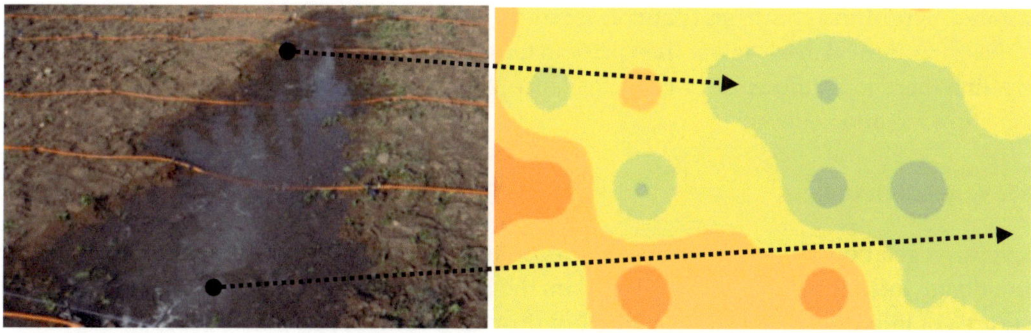

Figure 8. Detail of Figure 4. The lowest temperature values were recorded in the area were the leakage was still on going. The temperature map was able to draw with relative high precision the problematic area

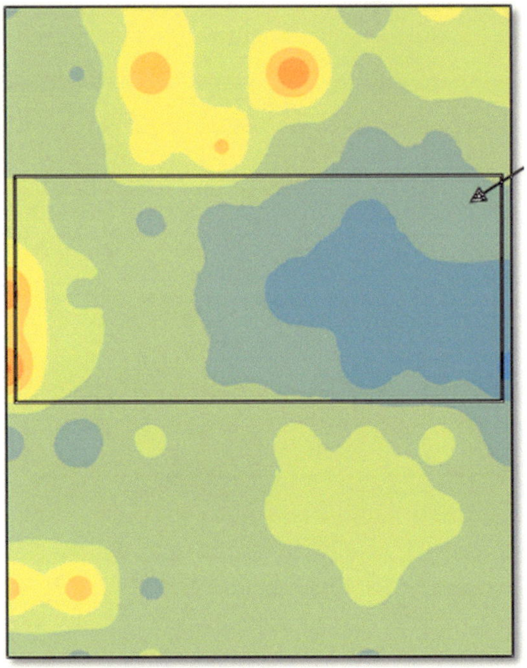

Figure 9. Ground temperature over the area of interest as recorded from the digital laser thermometer, during the leakage event. Rectangle indicates the location of the buried pipeline. Blue values indicates temperatures 14-16°C, 24-26°C are shown in yellow while orange colour highlight values from 34-36°C

4.2. Spectral signatures profiles

Figure 10 features the spectral signatures profiles from some regions of the study area before and after the leakage event. Using this figure, some preliminary conclusions can be drawn in relation with the reflectivity values of the same regions prior and after the leakage. The near infrared part of the spectrum (Band 4) tends to give the highest deviations. The reflectance values after the leakage is lower than prior to the event. Regarding the measurements over the pipeline there is a difference in the reflectance values before and after the leakage (spectral signature red and yellow).

4.3. Vegetation indices

Several vegetation indices were used in this study as shown in Table 1. Fifteen different broadband indices were calculated based on the RSR filters of Landsat 7 ETM+and applied in the area of interest. Some indices tended to provide poor results in contrast to other indices, which were able to detect the leakage. Several figures were created showing the area prior and after the leakage, as well maps showing the vegetation index difference.

As indicated in Figure 11, the EVI index was able to detect the leakage occurred over the buried pipeline with great success. As shown in figure 11, the highest negative values (blue colour)

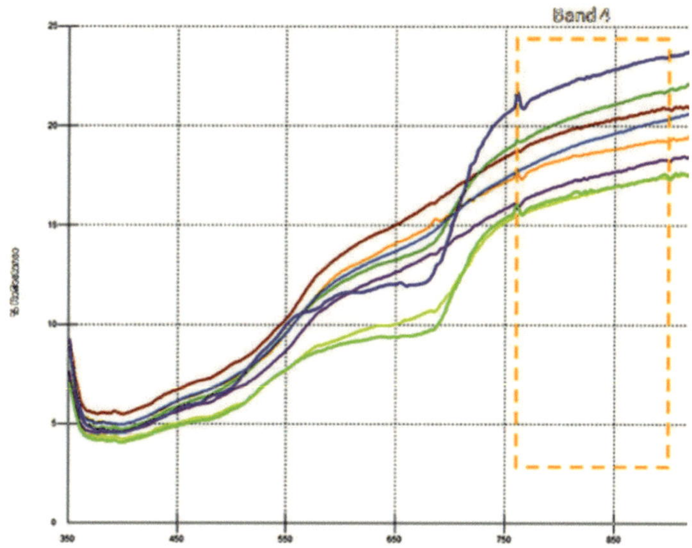

Figure 10. Spectral signatures profiles before and after the leakage event. The differences are maximized in the near infrared part of the spectrum (Band 4)

are observed in the area over the pipeline. In order to create this map, the EVI index was calculated before and after the leakage event and then these two indexes were subtracted. In a similar manner, the RDVI index was also able to show this leakage. A cross section from North to South was performed to see the vertical profile of the index. As demonstrated in Figure 12, the highest difference is recorded as the section pass through the pipeline (points A and B in Figure 12).

Several vegetation indices presented in Figures 13-25 were able to highlight the leakage as well as the area around the buried pipeline before and after the leakage (table 2). The buried pipeline is indicted by the black triangle in each figure. However, in addition to the leakage event, there were other parameters that played a significant role for the detection of the problem. As the results indicate, vegetation indices, including the NDVI index, were able to identify the leakage, yet there were some problems related to the atmospheric conditions. In contrast, the ARVI index was able to map the leakage with no difficulties since it is an atmospheric resistance index. EVI was another index with can be applied for this purposes as well.

Figure	Vegetation indice used
Figure 13	SRxNDVI
Figure 14	SR
Figure 15	Sarvi

Figure	Vegetation indice used
Figure 16	RVI
Figure 17	OSAVI
Figure 18	NDVI
Figure 19	MSR
Figure 20	MSAVI
Figure 21	IRG
Figure 22	GNDVI
Figure 23	GEMI
Figure 24	DVI
Figure 25	ARVI

Table 2. Vegetation indices used in this study, by figure

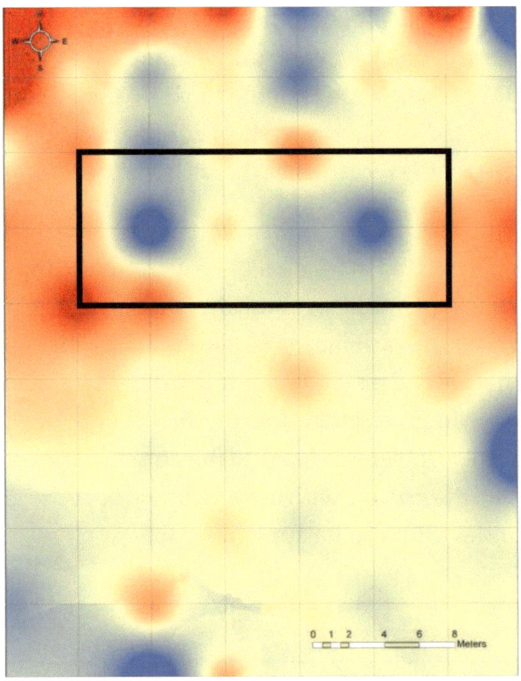

Figure 11. Difference values as calculated for EVI index (before and after the leakage). The buried pipeline is indicated in the rectangle.

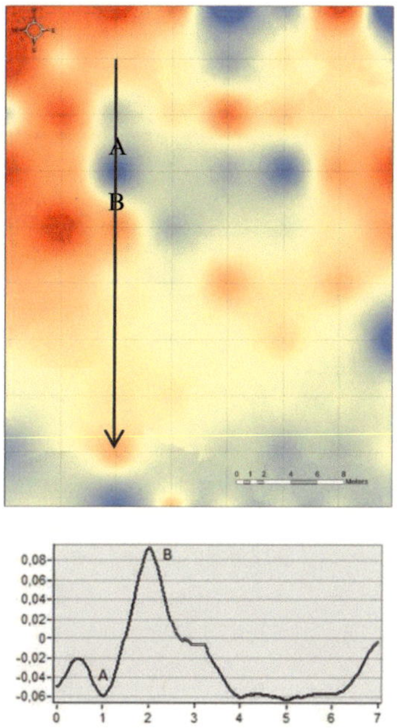

Figure 12. Difference values as calculated for RDVI index (before and after the leakage). A cross section passing through the pipeline (A-B) is also drawn.

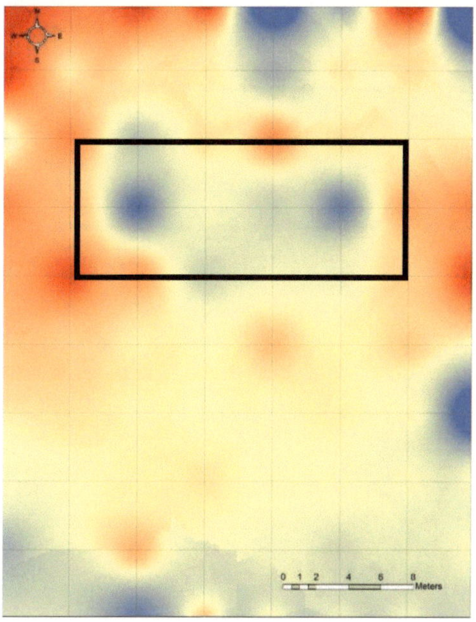

Figure 13. Difference values as calculated for SRxNDVI index

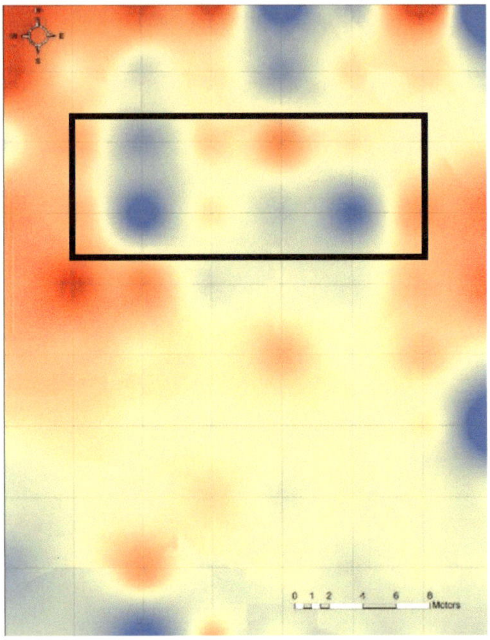

Figure 14. Difference values as calculated for SR index

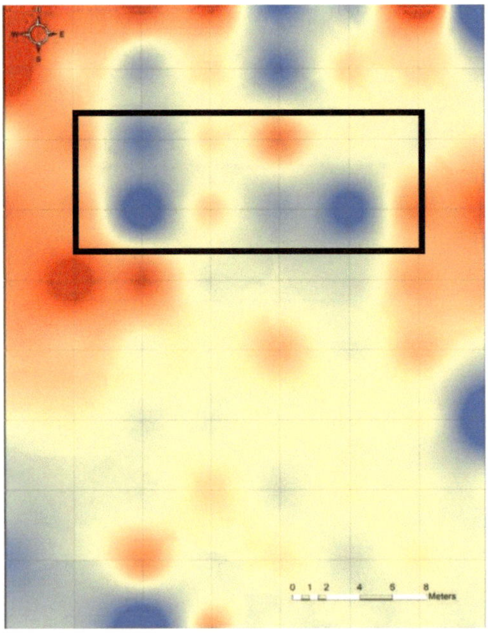

Figure 15. Difference values as calculated for SARVI index

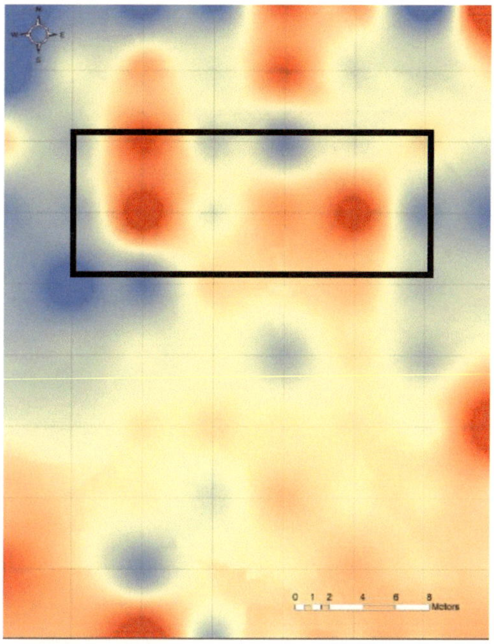

Figure 16. Difference values as calculated for RVI index

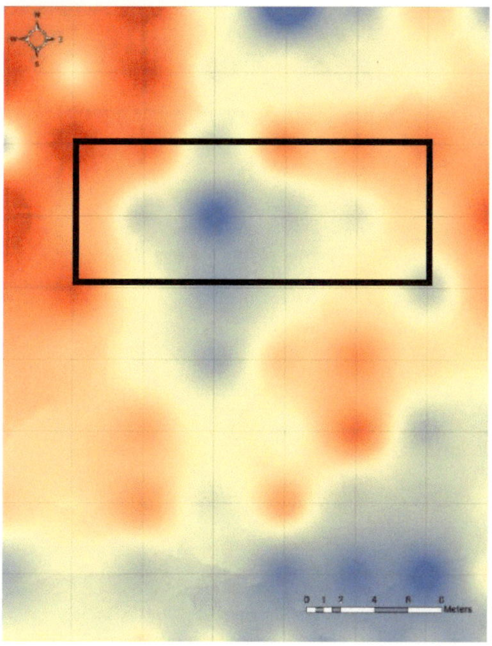

Figure 17. Difference values as calculated for OSAVI index

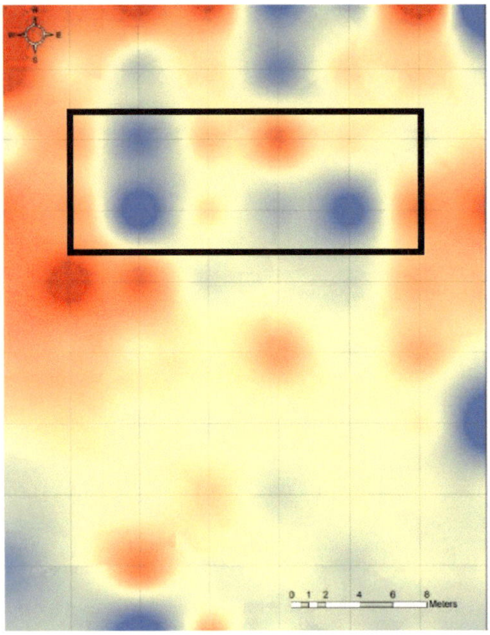

Figure 18. Difference values as calculated for NDVI index

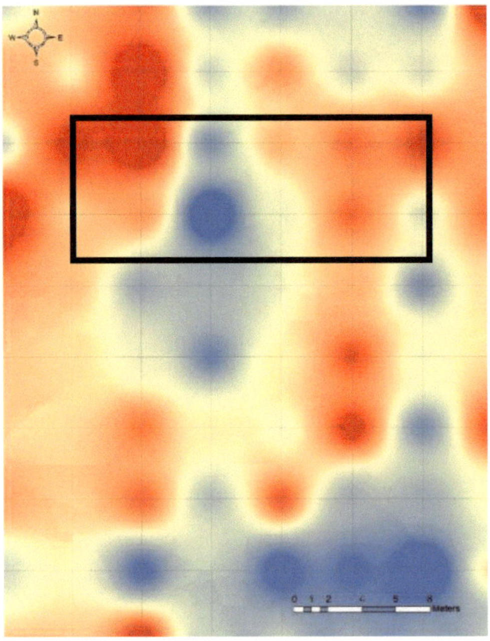

Figure 19. Difference values as calculated for MSR index

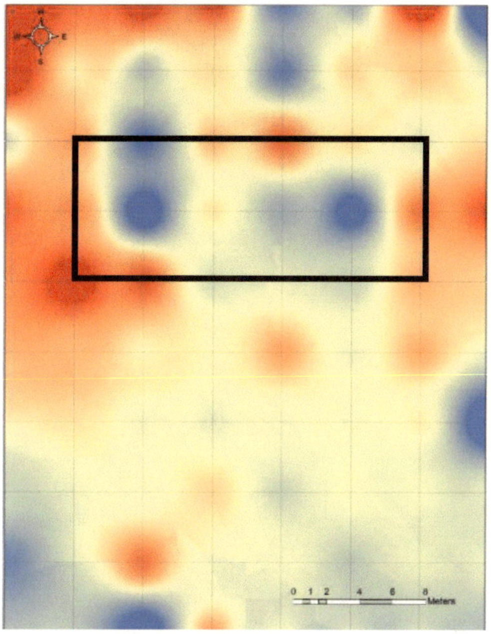

Figure 20. Difference values as calculated for MSAVI index

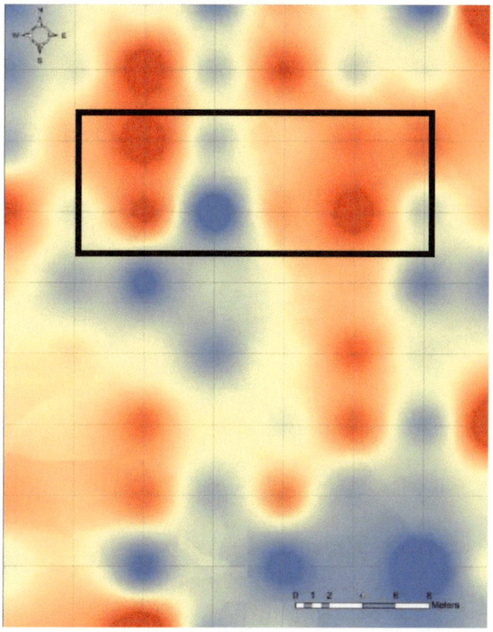

Figure 21. Difference values as calculated for IRG index

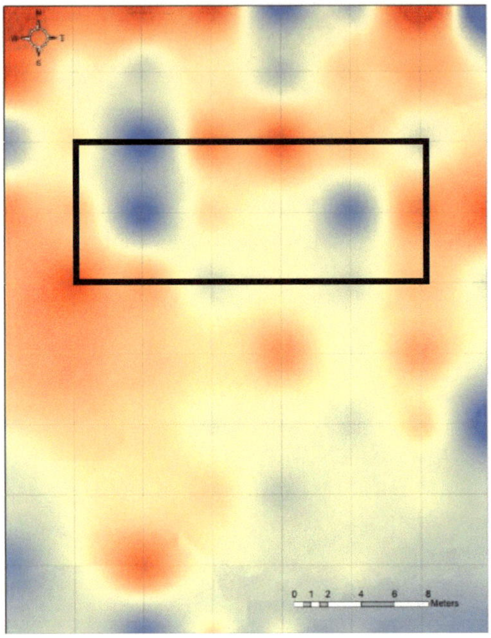

Figure 22. Difference values as calculated for GNDVI index

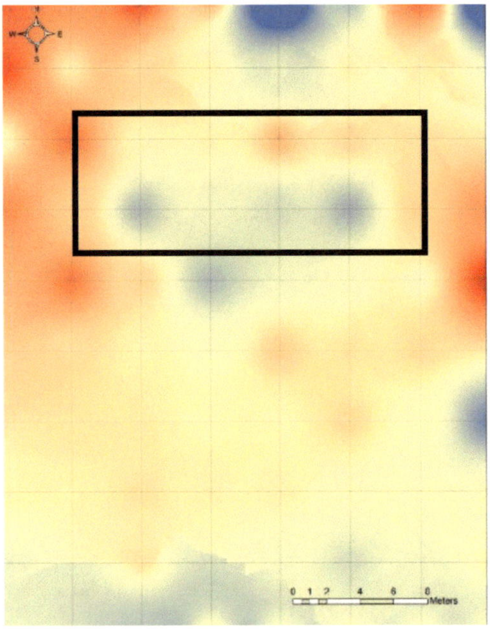

Figure 23. Difference values as calculated for GEMI index

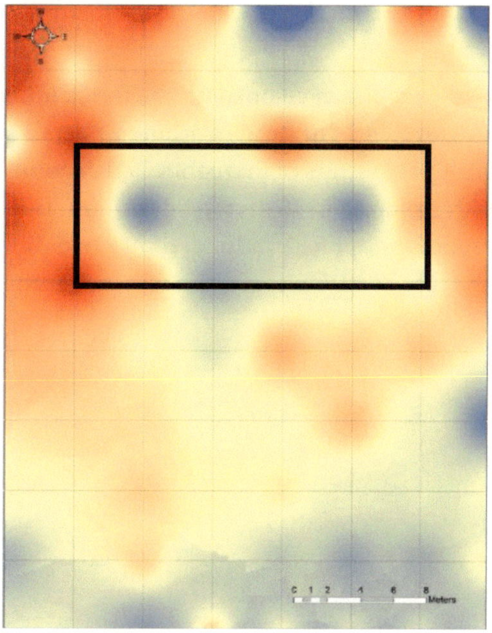

Figure 24. Difference values as calculated for DVI index

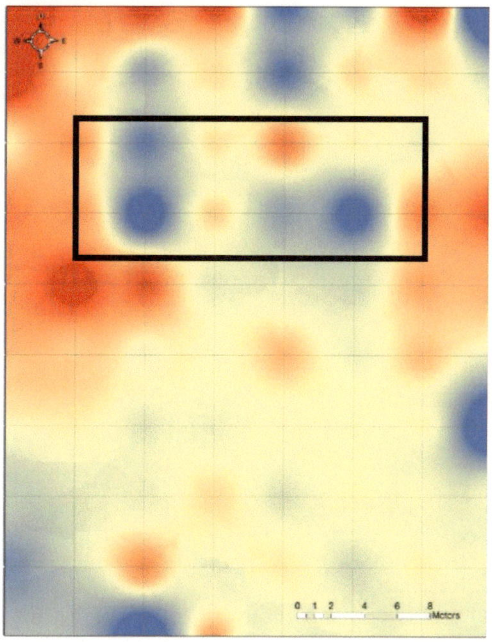

Figure 25. Difference values as calculated for ARVI index

5. Development of a leakage early warning system

During the study, a methodology was developed for the local authorities and other end users regarding the early warning system. The early warning system was based on the study by Agapiou et al. (2014) and from the results of this controlled experiment. The system was built in the ArcGIS software with the objective to extract useful information from remote sensing satellite images. The methodology applied is divided into two different scenarios according to the amount of satellite images available to end users (see Figure 26).

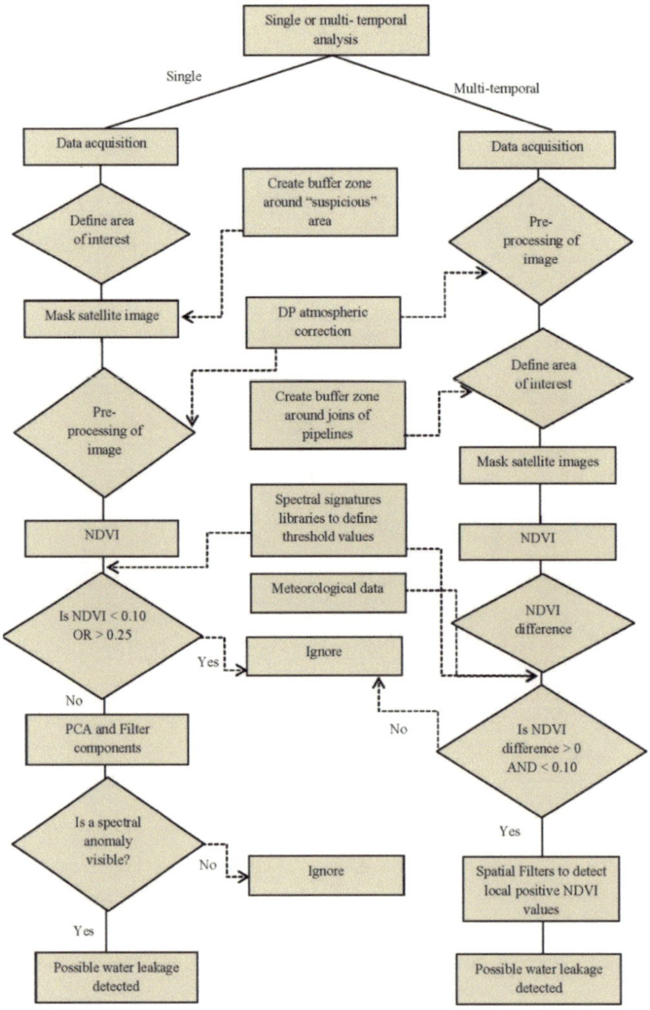

Figure 26. Methodology followed for the early warning system (Agapiou et al. 2014).

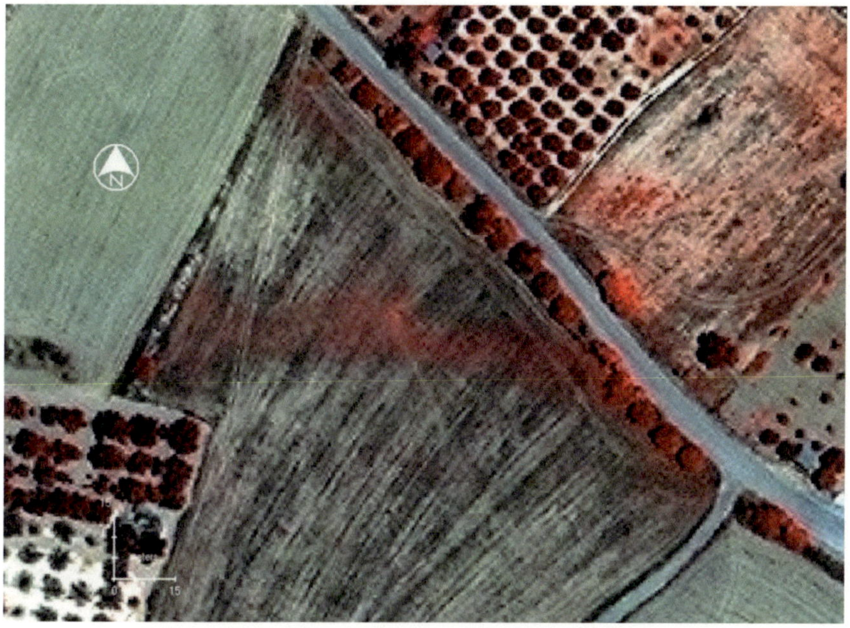

Figure 27. Detection of leakage problem using high resolution satellite data (Agapiou et al. 2014)

The first scenario aims to detect "suspicious areas" and therefore validate a hypothesis of a water-leakage event while the second scenario aims to detect unknown leakages events.

The first scenario is briefly outlined by the next steps:

Define the broader zone of "suspicious" leakages areas

Radiometric and Geometric corrections of the image, including atmospheric correction. In this study, the Darkest Pixel image based algorithm was applied since is a fast and accurate method for removing atmospheric effects

Calculation of the NDVI index in the area of interest

Definition of thresholds for the NDVI values. These thresholds were defined a-priori after field spectroradiometric campaigns

Masking of areas within thresholds boundaries

Application of other vegetation indices and PCA analysis as well as spatial filters for the detection of vegetated areas

Definition of areas with possible leakage problem

For the second scenario the steps are:

Radiometric and Geometric corrections of the images

Creation of buffer zones around the joints of the water pipeline

Masking of images

Calculation of NDVI index in the area of interest in the whole dataset

Calculation of NDVI difference between the two images

Definition of thresholds for the NDVI values based on pre-defined thresholds.

Definition of areas with possible leakage problems using spatial filters

The modeler was built into the ArcGIS environment and successfully applied using several satellite data. Previous studies of the authors (Agapiou et al. 2014) the detection of leakage events was possible (see Figure 27).

6. Discussion

Leakage detection is very important for water management. As presented in this chapter, research has indicated that remote sensing techniques can identify leakage detection. Remote sensing techniques can provide useful data, both for the detection of the water pipes and for the detection of water leakages. This study was conducted in a controlled environment using ground remote sensing techniques. The use of ground spectroradiometers has shown that several broadband vegetation indices may be applied for the detection of leakages events. The different maps created in this chapter indicate that some indices such as ARVI or EVI are favourable for this purpose. The results demonstrated that digital thermometers may be used for the detection of leakages events. Therefore, either airborne or satellite systems equipped with thermal sensors and with a high resolution may be possible detect areas with leakage problems. As well, UAVs can be used successfully to identify water leakages, through the use of aerial photography and other sensors. The results of this study indicate that remote sensing techniques are able to detect areas of the pipeline with water leakages. The resulting data can be integrated into a GIS which can be used by local authorities as an early warning system.

Acknowledgements

The results reported here are based on findings of the Cyprus Research Promotion Foundation project "ΑΕΙΦΟΡΙΑ/ΦΥΣΗ/0311(BIE)/21": Integrated use of space, geophysical and hyper-spectral technologies intended for monitoring water leakages in water supply networks in Cyprus. The project is co-funded by the Republic of Cyprus and the European Regional Development Funds of the EU. Thanks are also given to the Remote Sensing and Geo-Environment Laboratory of the Department of Civil Engineering & Geomatics at the Cyprus University of Technology for its continuous support (http://www.cut.ac.cy).

http://dx.doi.org/10.5772/59531

Author details

Athos Agapiou[1*], Dimitrios D. Alexakis[1], Kyriacos Themistocleous[1], Apostolos Sarris[2], Skevi Perdikou[3], Chris Clayton[4] and Diofantos G. Hadjimitsis[1]

*Address all correspondence to: athos.agapiou@cut.ac.cy

1 Department of Civil Engineering and Geomatics, Faculty of Engineering and Technology, Cyprus University of Technology, Cyprus

2 Laboratory of Geophysical - Satellite Remote Sensing & Archaeo-environment (GeoSat ReseArch), Institute for Mediterranean Studies (I.M.S), Foundation for Research & Technology, Hellas (F.O.R.T.H.), Crete, Greece

3 Department of Civil Engineering and Geomatics, Faculty of Civil Engineering, Frederick University, Cyprus

4 University of Southampton, School of Engineering and the Environment, UK

References

[1] Agapiou, A., Alexakis, D.D., Themistocleous, K. and Hadjimitsis. Water leakages detection using remote sensing, field specstroscopy and GIS in semiarid areas of Cyprus. *Urban Water Journal*, In press.

[2] AWWA. *Leaks in Water Distribution Systems – A Technical/Economic Overview*. Denver, Colorado: American Water Works Assoication. 1987.

[3] Cheong, L.C.. Unaccounted-for Water and the Economics of leak detection, in:Water Supply (eds.). *Proceedings of the International Water Supply Congress and Exhibition*. Copenhagen. Copenhagen: Water Supply. 1991. 1-6.

[4] Cho, G., (et al.) 2013. Pilotless aerial vehicle systems: size, scale and functions. *Coordinates*, 9, pp.8-16.

[5] Colomina, I. and Molina, P. 2014. Unmanned aerial systems for photogrammetry and remote sensing: a review. *ISPRS Journal of Photogrammetry and Remote Sensing*, 92, pp. 79-97.

[6] Huang, Y and Fipps, G. *Thermal Imaging of Canals for Remote Detection of Leaks: Evaluation in the United Irrigation District*-Technical Report. A&M University Texas: Biological and Agricultural Enginnering Department. 2002.

[7] Hunaidi, O. and Chu, W.T.. Acoustical characteritsics of leak signals in plastic water distribution pipes. *Applied Acoustics*, 1999. 58(235-254)

[8] Hunaidi, O and Giamou, P.. Ground Penetrating Radar for detection of leaks in buried plastic water distribution pipes, in:*Proceedings of the Seventh International Conference on ground Penetrating Radar*. Lawrence, Kansas. Lawrence, Kansas: 1998. 27-30.

[9] Mayr, W., 2013. Unmanned aerial systems-for the rest of us. Proc. 53rd Photogrammetric Week. Institute fur Photogrammetrie, Universitat Stuttgart, pp.125-134, (2013).

[10] Petrie, G., 2013. Commercial operation of lightweight UAVs for aerial imaging and mapping. GEOInformatics, 16, pp.28-39.

[11] Pickerill, J.M. and Malthus, T.J.. Leak Detection from Rural Aqueducts using Airborne Remote Sensing Techniques. *International Journal of Remote Sensing,* 1998. 19(12): pp. 2427-2433.

[12] Shih, S.F. and Jordan, J.D.. Use of Landsat Thermal-IR data and GIS in soil moisture assessment. *Journal of Irrigation and Drainage Engineering,* 1993. 119(5): pp. 868-879.

Chapter 3

Monitoring of Water Leakages from Pipes through Geophysical Imaging Methods

Meropi Manataki, Nikos Papadopoulos,
Apostolos Sarris, Athos Agapiou,
Kyriacos Themistocleous and
Diofantos G. Hadjimitsis

Additional information is available at the end of the chapter

http://dx.doi.org/10.5772/59532

1. Introduction

Electrical Resistance Tomography (ERT) and Ground Penetrating Radar (GPR) are among the near-surface geophysical methods that are suitable for mapping the subsurface. As with all the geophysical methods, the ability of ERT and GPR methods to provide images of buried structures relies on the contrast of soil properties. The former provides images of the subsurface by measuring the electrical resistivity, while the latter records the time arrivals of electromagnetic pulses that are affected mostly by the differences of soil electrical properties. Since water leaks cause changes in soil electrical properties, it is feasible for water leakages to be detected without the need or soil drilling.

In this chapter, the results of controlled experiments for monitoring leakages from water pipes are presented. ERT and GPR were both applied in a test field site at Lakatameia Municipality to monitor the change of subsurface physical properties due the infiltration of water within the soil matrix. Time-lapse mode for both techniques was used to monitor the flow progress of the water within the ground surface due to the controlled leakage from a water pipe. These geophysical techniques were chosen as the most appropriate for meeting the needs of the research, the geomorphological characteristics of the site and the expected subsurface archaeological targets with regards to the detection and mapping of the targets.

2. Electrical Resistivity Surveying principles

Resistivity surveying[1] is one of the most commonly applied technique of geophysical survey in the Mediterranean due to its suitability in detecting walls, cavities, layers and other localized structures of differing electrical, permittivity and electrochemical properties (Mares, 1984). The electrical resistivity method has been applied with great success in solving hydrogeological (Dahlin and Owen, 1998), geological (Atzemoglou et al., 2003), engineering (Dahlin et al., 1994), mineral (Legault et al., 2008), environmental (Ramirez et al., 1996), archaeological survey problems (Papadopoulos et al., 2006) and detecting leaks from buried pipes (Jordana et al., 2001).

From the methodological point of view, in geoelectric surveys, a direct current or an alternating current of low frequency (<100 Hz) is fed into the ground by two metallic current electrodes with low stake resistance. This causes a potential field, which is influenced by the distribution of the specific resistivities in the earth. Measurements of differences of the potential field are carried out between two well-grounded, non-polarizable potential electrodes configuration. By applying special evaluation software, information about the distribution of resistivities and the geological structures can be derived. The term "apparent resistivity" is frequently used in the interpretation of resistivity data in order to overcome difficulties arising from inhomogeneous ground.

Electrode array or configuration refers to the specific way that the four electrodes are placed on or inside the ground in order to complete a specific resistivity measurement. Linear arrays, where the electrodes are placed along a line, are the most common due to the relative simplicity of the measured signals. The selection of the appropriate array for a given survey depends on diverse factors like the size of the survey, the nature and depth of the targets, the properties of the subsoil, the type of equipment, the depth of investigation and the array's behaviour characteristics (Weymouth and Huggins, 1985).

The method is normally employed in mapping mode employing a fixed electrode configuration moving along multiple parallel profiles forming a grid giving the horizontal distribution of subsurface resistance in a constant depth. This technique is very common in the exploration of archaeological sites mapping cultural and architectural buried relics (Aspinall and Lynam, 1970). Multi-probe, wheeled and tractor-based systems have also been able to speed up resistivity surveying and cover extensive areas mainly in smooth and conductive soils (Panissod et al., 1998).

The fixed electrode separation imposes specific limits for the maximum investigation depth that are closely related to the distance between the electrodes. The current flows along arc-shaped paths that connect the two electrodes and, by increasing the electrode spacing, more of the injected current will flow to greater depths. Thus, when depth information is needed, the vertical electrical sounding can measure the vertical distribution of the resistance (i.e. the

1 Vertical Resistivity Soundings (VES), Resistivity Profiles (LP) and Electrical Resistivity Tomography (ERT)

depth of a feature) by increasing the electrode separation and at the same time keeping the center of the configuration stable.

By combining along a line multiple profiles, with increasing electrode distances or alternatively multiple soundings with different centers, it is possible to record the vertical and horizontal distributions of resistivity along transects producing a two-dimensional (2-D) vertical resistivity section of the subsurface. In the cases of complex geology, artefacts may be generated in the two dimensional sections due to the three-dimensional (3-D) nature of the underground structures. This problem can only be resolved with the collection of full 3-D measurements where a finite number of electrodes are placed on the nodes of a rectangular grid and the potentials caused by a single current injection are measured at the remaining electrodes in all the different directions.

However, due to the logistics, full 3-D surveys cannot be applied on a routine basis. Therefore, even today the most common practice for recording the 3-D subsurface resistivity variation is to gather dense, parallel 2-D lines with the inter-line spacing equal to the basic inter-electrode spacing along a single or orthogonal directions (Papadopoulos et al., 2007). In the past 25 years, the development of the technology related to multiplexed electrode arrangements, automatic measuring systems and processing algorithms have resulted in the evolvement of Electrical Resistivity Tomography (ERT) technique (Loke et al., 2013).

The arrival of fast and powerful computers, in combination to the non-linear and ill-condition nature of the resistivity problem, has generated the necessity of developing automated modeling and inversion algorithms to process and interpret the resistivity data. These algorithms involve the reconstruction of the subsurface resistivity distribution which is consistent with the experimental data through an iterative optimization procedure that imposes smoothness or blocky constraints in order to stabilize the inversion procedure and produce a reliable resistivity model (Loke and Barker, 1996, Loke et al., 2003, Papadopoulos et al., 2011).

ERT can also be used to monitor dynamic processes that evolve in the ground by installing permanent electrodes in fixed positions. In such cases, time lapse resistivity surveys or 4-D imaging techniques (Karaoulis et al., 2011; Kim et al., 2009) approach the spatial changes of the subsurface resistivity distribution over time. In general these resistivity changes can only be mapped and monitored only if substantial spatial-temporal changes of the measured apparent resistivity exist. Typical examples of such approaches are the flow of water through the vadose zone, the flow of chemical pollutants, the vertical variations of the water table, etc. A number of different approaches have been proposed for the efficient processing and interpretation of 4-D imaging ERT data (e.g. independent inversions, ratio of inverted data from initial and later data sets, difference inversions, and least-squares regularization).

3. Ground Penetrating Radar (GPR) principles

Ground Penetrating Radar (GPR) is a non-destructive electromagnetic (EM) geophysical technique that uses radio waves, in the frequency range of 10MHz to 2GHz, to map the subsurface. The first reported attempt of using radio wave signals to measure subsurface features was by El-Said, (1956) who tried to image the depth of a water table. The development of the method accelerated considerably after 1970 as a result of the tremendous progress that took place in electronics and computer technology (Annan, 2002).

GPR can be used in a series of applications like the mapping of the bedrock depth (Davis and Annan, 1989), the determination of the stratum thickness and the aquifer depth (Doolittle et al., 2006), the location of physical and artificial cavities in the subsurface (Benson, 1995) and fracture zones (Grasmueck, 1996; Theune et al., 2006). Recently the method is also used in archaeological prospection for the detection of buried antiquities (Papadopoulos et al., 2009; Spanoudakis et al., 2011; Vaughan, 1986). Additionally, GPR has been used for locating both buried pipes and water leakages providing promising results (Bimpas et al., 2010; Hunaidi et al., 1999; Powers and Olhoeft, 1996; Stampolidis et al., 2003).

The operation principle of GPR lies on the interaction between the EM energy and the matter, and has specific similarities with the seismic reflection method. High frequency pulses of short duration are injected into the ground by a transmitter antenna. Those pulses or signals are "traveling" into the subsurface until they hit a boundary of a different material. Part of the energy is reflected back to the surface and recorded by the receiver antenna, while the remaining energy is diffused at deeper levels. This boundary or reflector is defined by the contrast on the electrical properties of the subsoil's materials, the conductivity and permittivity, which affect the wave's propagation. The receiver records the amplitude of the reflected signal with respect to the travel time, i.e. the time that the signal needed to reach the receiver from the transmitter. Basically the method depends on the record of the waves reflected on surfaces that divide regions with different electrical properties.

The time between the transmitting and the receiving pulse depends on the velocity along the trace the pulse followed. The paths that an emitted signal can follow in the case of two layers with different electrical properties are illustrated in Figure 1. Since the time is measured, if the electromagnetic wave propagation velocity is known, then the depth of the reflector can be determined. In most geological materials, the conductivity and the dielectric constant (relative permittivity) are the main parameters that affect the pulse. Conductivity affects the energy absorption, thus the signal penetration and the depth of the investigation, while permittivity affects the velocity of the signal. In general, GPR is most useful in low-electrical-loss materials (i.e. very low conductivity values). Clay-rich environments or areas of saline water will affect negatively the method's efficiency (Cassidy, 2009a).

Besides the electrical properties of subsoil's materials, the penetration depth of GPR depends also from the central frequency of the emitted pulses. The absorption increases with frequency and thus a smaller frequency is used for detecting the deeper targets. However, the resolution of the method is decreased as the frequency is increased. For instance, a GPR system working at the range of 25-50 MHz can investigate depths over 50 meters in soils with low conductivity (smaller than 1ns/m) like sand and gravels.

The data are collected by moving the configuration of transmitter and receiver (common offset) along a profile on the surface. The recordings produce a 2D image of the subsurface (Distance

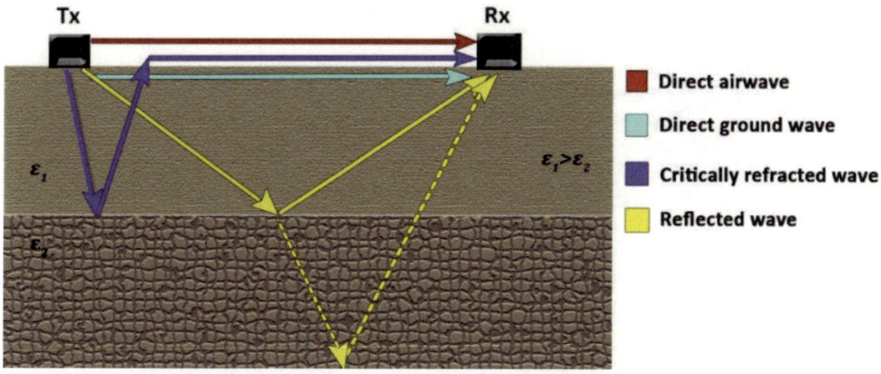

Figure 1. Signal paths between the transmitter (Tx) and the receiver (Rx) for two layers with different electrical properties. The dashed yellow line indicates that part of the signal's energy was refracted and then reflected (Annan, 2009).

(m)-Time (ns) or Depth (m)) that is called radargram or section (Fig. 2a). When working in grids of constant line spacing, 3D images of the subsurface are obtained out of which depth slices can be are extracted (Fig. 2b). Depth slices are also 2D images that provide the horizontal information of the subsoil with increasing depths.

Data processing of GPR data can be divided into two major stages: (1) the processing of the radargrams where signal processing techniques are used and slices are extracted and (2) the processing of slices where image processing corrections are applied. More emphasis is given to the first stage that aims to filter the noise from the data and enhance the reflections of the raw data. Some standard processes that are usually applied regardless the field of application, are (Cassidy, 2009b):

Traces reposition that corrects the position of GPR traces included in a survey line. This correction is useful to eliminate systematically offsets in survey lines' starting and ending positions that is usually occurred in rough terrains.

Timezero correction which allows to estimate the correct vertical position of the first pulse that left the antenna and entered the subsurface (Tzanis, 2006).

Dewow filter which removes low frequency noise derived by low frequency energy near transmitter and is associated with electrostatic and inductive fields.

SEC (Spreading & Exponential Compensation) gain that enhance signals located at greater depths and have much smaller amplitude compared with that of the shallower signals.

Background subtraction filter: that reduces random noise from the data and also removes the direct waves and ringing noise.

Hilbert Transform in order to calculate the instantaneous amplitude (Spanoudakis and Vafidis, 2010).

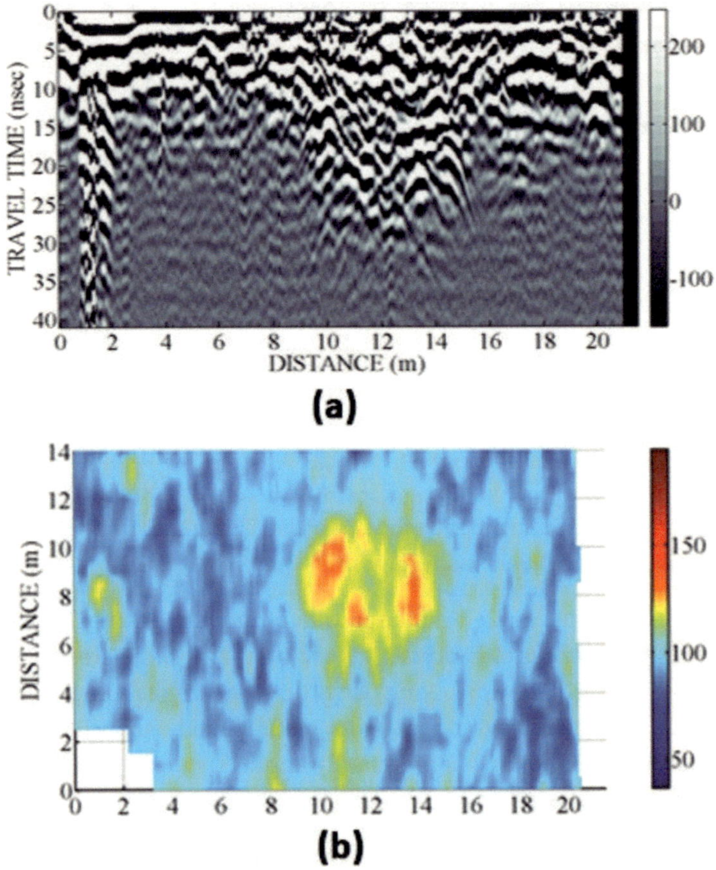

Figure 2. Images of buried roman cistern located at Aptera archaeological site where (a) is a GPR section while (b) is a GPR slice revealing the shape of the cistern at the depth of 1,0 m. PulseEKKO 1000 with a 400MHz was employed with 0.5m spacing between the transects (Spanoudakis et al., 2011)

4. Case study: Lakatameia Municipality

The test site is located at Lakatameia Municipality of Cyprus (Fig. 3a). An area of 420 square meters (15 x 28 m) within the green dashed polygon (Fig. 3b) was surveyed and monitored with both ERT and GPR. Water was injected to the ground through pipe 1 (Fig. 3b) to create a controlled water leakage located approximately at 2-3m depth. Time lapse mode was used for both methods. The geophysical monitoring mapping survey of the site was conducted for a period of two days in the beginning of summer of 2013.

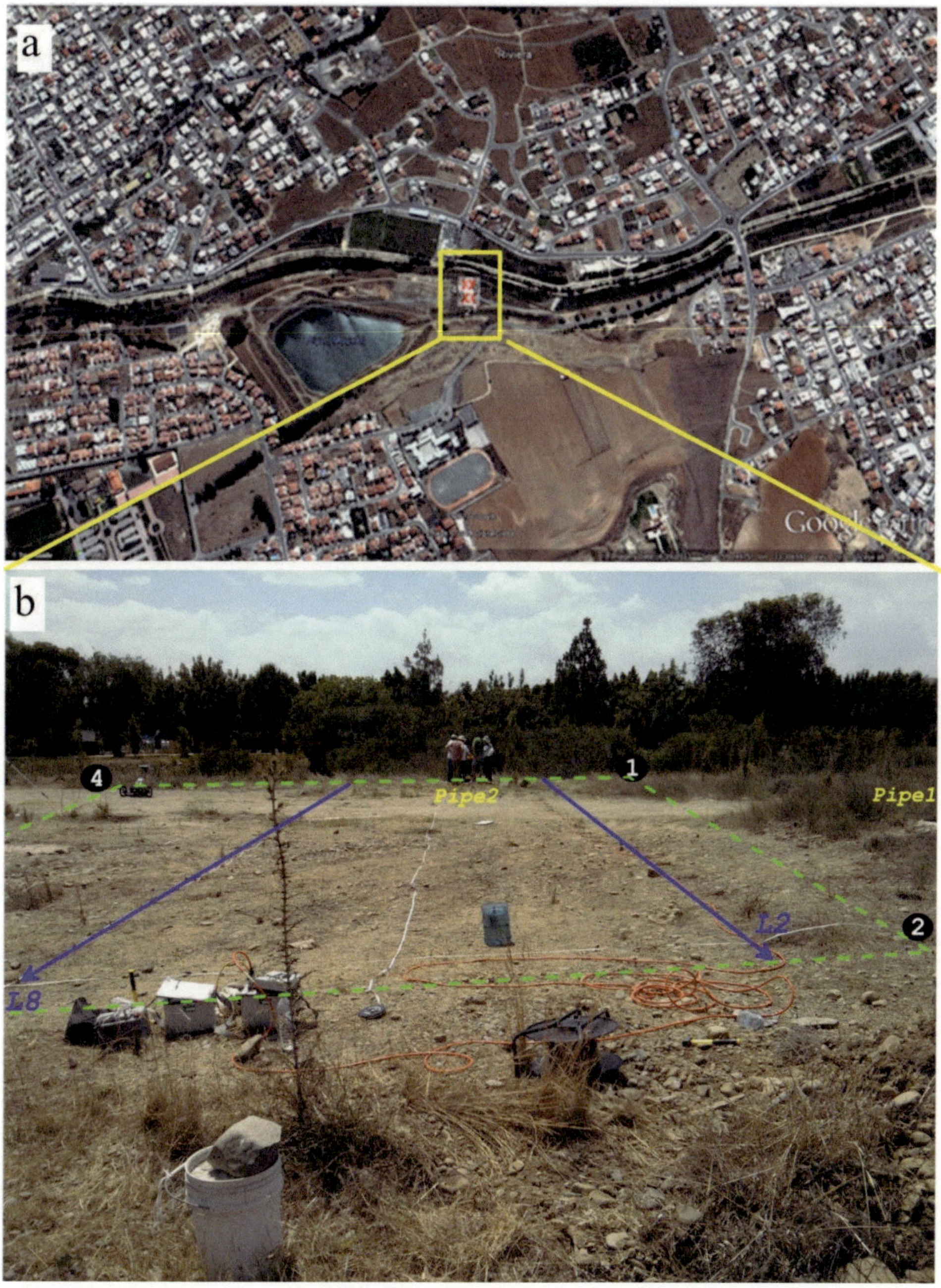

Figure 3. Google Earth satellite image of Lakatameia Municipality (Cyprus) showing the area where the monitoring experiment was conducted. b) The dashed polygon outline the geophysical survey area. The blue arrows indicate the direction of the surface ERT lines that were used to monitor the water leakage that was initiated from Pipe 1.

4.1. Instrumentation and Data acquisition

A specific map coordinate system was chosen for the geophysical grid, which was registered to the appropriate geodetic system of coordinates (local coordinate system), based on the Total Station mapping data. Thus, after the rectification of the satellite image, the geophysical maps can be overlaid at their corresponding location.

For ERT, the Iris Syscal Pro and Switch Pro 96s was used (Fig. 4a) while for the GPR, the Noggin Plus Smart Cart was employed, equipped with a 250MHz antenna (Fig. 4b). The expecting investigation depth is 5.0-6.0m for ERT while 2.0-3.0m for GPR. The geophysical survey parameters for both methods are summarized in Table 1, while the data acquisition is described in the following paragraphs.

Figure 4. Details of the geophysical instrumentation that was used in the prospection at Lakatamia Municipality where a) ERT instrumentation and (b) GPR system equipped with 250MHZ antenna.

Method	Δx (m)	Δy (m)	Area Coverage
GPR *Monitoring survey, 13 phases*	0.5, 1.0	0.025	420 m^2
ERT *static survey,* *8 parallel ERT lines, Dipole-Dipole array* *29 electrodes in each line, Length of lines; 28m*	1.0	2	392 m^2
ERT *monitoring survey,* *2 lines (L2 & L8), Dipole-Dipole array* *29 electrodes in each line, Length of lines; 28m* *9 Monitoring phases*	1.0		

Table 1. Technical details of the geophysical survey parameters.

4.2. ERT geophysical survey

The first day was devoted to completing a static 3-D ERT survey by collecting 9 parallel ERT lines (L1 to L9). Each line was 28 meters long and 29 electrodes were placed along each transect in equally spaced intervals (a=1m). The inter-line separation was 2 meters covering in this way the whole area of interest. The Dipole-Dipole electrode array was chosen to capture the apparent resistivity measurements using the same configuration protocol for all the individual lines (a=1m, 1a: Nsep=1a-8a, 2a: Nsep=1/2a-15/2a & 1a-8a, 3a: Nsep=8/3a-16/3a). The goal of this survey was to map the stratigraphy and compile a model of the site regarding the main lithological layers up to the depth of about 5-6 meters below the ground surface.

The monitoring survey took place during the second day. The controlled water leakage from the pipe 1 (Fig. 3b) was initiated at 10:10 am and terminated four and a half hours later at 14:40 hours (Table 2). The reference ERT measurements for Lines L2 and L8 (Fig. 3b) were acquired at 10:11am and 10:44am respectively. The protocol which was used for the static ERT survey was also implemented for both lines and all the time phases. Thus the actual time for the data acquisition along each line was the same for all the snapshots that were taken (7-8 minutes).

	L2			L8		
	Leakage Initiation: 10:10 (June 13th, 2013)					
Phase	*start*	*end*	*duration*	*start*	*end*	*duration*
T1	10:11	10:19	00:08	10:44	10:51	00:07
T2	10:56	11:04	00:08	11:12	11:19	00:07
T3	11:24	11:31	00:07	11:41	11:48	00:07
T4	11:56	12:03	00:07	12:12	12:20	00:08
T5	12:24	12:31	00:07	12:36	12:43	00:07
T6	12:50	12:57	00:07	13:03	13:11	00:08
T7	13:19	13:27	00:08	13:43	13:50	00:07
T8	14:08	14:15	00:07	14:21	14:28	00:07
T9	14:33	14:40	00:07	14:44	14:52	00:08
	End of ERT Monitoring: 14:52 (June 13th, 2013)					

Table 2. Details regarding the initiation/termination time of the water leakage initiated the starting/ending point for 9 different time phases that were used to monitor the leakage.

4.3. GPR geophysical survey

GPR was employed on the second day, simultaneously with ERT. The first grid (GRID0) was measured before the beginning of the water leakage. Parallel lines were used with 0.5m spacing, starting from point 1 while the orientation was the same with ERT lines (Fig. 5). The rest of the grids (12 in total) were collected in different times after the initiation of the leakage. The line spacing for the monitoring grids is 1.0m and the starting / ending point was at 4.0m and 12.0 m respectively. The challenge with GPR is to test whether it is capable of providing information about the water leakage without including the pipe and check its effectiveness to monitor water leakages directly from the subsoil.

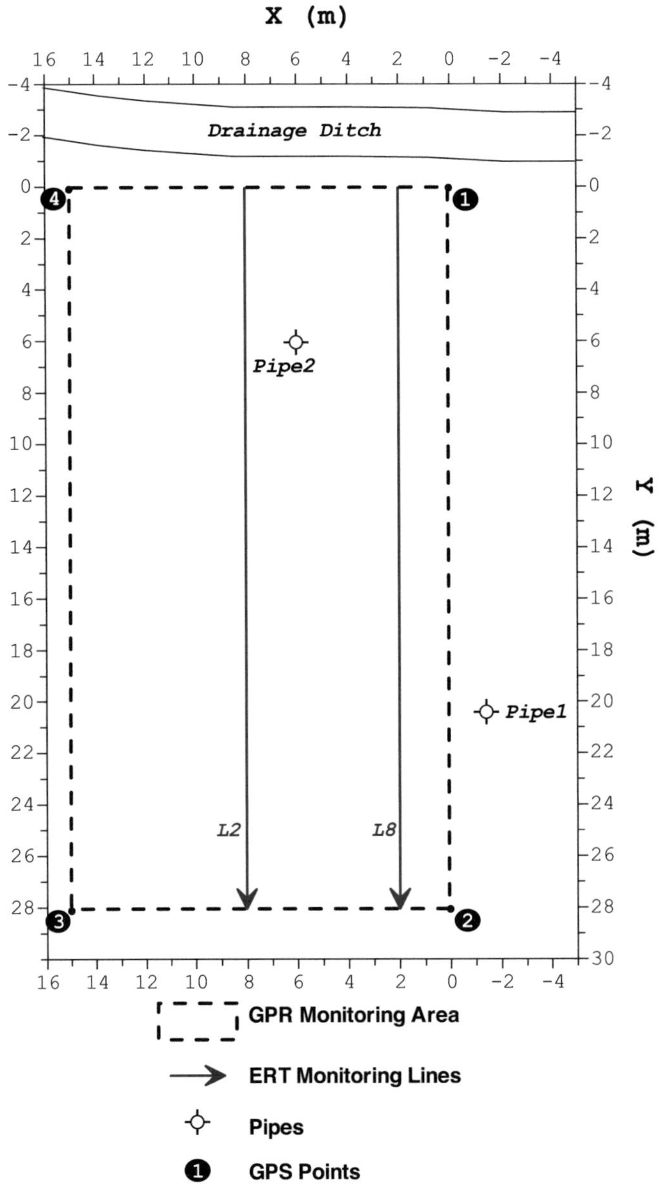

Figure 5. Area coverage of the GPR-ERT monitoring.

5. Data processing

A specific workflow was used to process the ERT and GPR geophysical data. The raw geophysical data were entered in a portable PC right after fieldwork. ERT data processing is divided into two phases, the static survey and monitoring survey. GPR filters and corrections were initially applied on each section in order to improve the signal to noise ratio and enhance the reflections. Following, the differences between the first (dry soil) and the second phase (start of the water leakage) were calculated to examine the changes in the subsoil due to the water flow.

5.1. ERT data processing

The static resistivity data were processed with a 3-D resistivity inversion algorithm (Papadopoulos, et al., 2011) that converged to a final resistivity model after 5 iterations and RMS=3.8%. A generalized and systematic way was used to process the time lapse ERT data that collected along the lines L2 and L8. The extreme noisy apparent resistivity measurements were removed through specific despiking filters. The time-lapse data were subjected to diverse processing methodologies in order to determine the most appropriate one. Within the time-lapse inversion procedure the following constraints were tested:

No constraints: The data for the time-lapse survey were processed independently. Then the changes of the subsurface resistivity values are determined by comparing the model resistivity values obtained from the inversions of an initial data and the later time data sets.

A damped least-squares constraint to minimize the differences in the model resistivity values between the initial model and the time-lapse model.

A Least-squares smoothness constraint to ensure that the differences in the model resistivity values vary in a smooth manner.

Robust constraints to minimize the absolute changes in the model resistivity values.

The tests with the various inversion techniques using different constraints showed comparative results resulting in smooth subsurface monitoring images. Robust constraints showed relatively sharper inversion models. After the completion of the time-lapse inversions, the percentage change in the model resistivity obtained from the inversion of a later time data set compared to a reference model was plotted for all the different time snapshots. For direct comparison of the relative resistivity change during the time phases, a common range was used for all the plots.

5.2. GPR data processing

The sections that derived from Sensors & Software Noggin Plus Smart Cart with 250 MHz antennas were processed using MATLAB and the packages GPR-Pro (Spanoudakis and Vafidis, 2010) and MAT-GPR (Tzanis, 2006). For each section, the following corrections were applied in order to enhance the signal:

Time zero correction in order to define the initial useful signal from each line. This determination was based on the intensity percentage of the first reflected wave (5-30%).

Dewow filter to remove low frequency noise derived from the antenna.

SEC gain to enhance reflections located in greater depths.

Background noise removal filter to remove ringing and random noise that gain function highlighted.

The processed sections for each grid were placed appropriately in a three dimensional array X (m), Y(m), Z(ns) according to the chosen local reference system that was used for the site, resulting in 13 volumes that represent the subsoil of the monitoring site for the different time lapses. Then the instantaneous envelope was estimated via Hilbert Transform and an average filter was applied for smoothing the results. The slices for GRID0 were extracted assuming a velocity for the electromagnetic waves equal to 0.1m/nsec, while for the rest of the grids the velocity was estimated from the existing hyperbolas because the presence of water changes the electromagnetic waves velocity.

Next, the following procedures took place in order to emphasize the water flow:

Procedure 1: Percentage differences using GRID0 as a reference.

Procedure 2: Percentage differences using GRID1 as a reference.

At the resulting volumes, slices were extracted with a step of 0.25m and colour maps were also applied.

6. Results

The 2-D resistivity section obtained from the static resistivity data (Fig. 6a) and extracted from the 3-D model shows generally a stratified earth with local superficial inhomogeneity. The range of the resistivity values between 50-300 Ohm-m signify a gravel/sandy layer from the surface up to the depth of 3m. The underlying horizon that exhibits lower resistivity signature (10-50 Ohm-m) is attributed to a silt and clay layer with sand. The low resistivity region along the section (X=7-12m / Z=0-1m) is related to a modern ditch that has been excavated during the installation of Pipe 2 in the area and it appears in all the parallel lines (Fig. 6b). The general geologic setting that was extracted from the 3-D ERT survey regarding the uppermost gravel/ sandy layer is in quite agreement with in situ observation of the sediments from a natural subsurface cross-section observed at the edge of site (Fig. 6c).

The reference ERT measurements for Line 2 began at 10:11am. Approximately one hour (11:04am – end of Phase T2) after the controlled leakage began, the subsurface resistivity properties started to exhibit fluctuations of approximately 6-7% decrease of resistivity. This decrease was registered along the horizontal distance X=4-8 meters and at a depth of more than 1.2 meters. As the monitoring of site continues, the resistivity decrease was more pronounced within the same region during the later phases. By the end of the experiment at

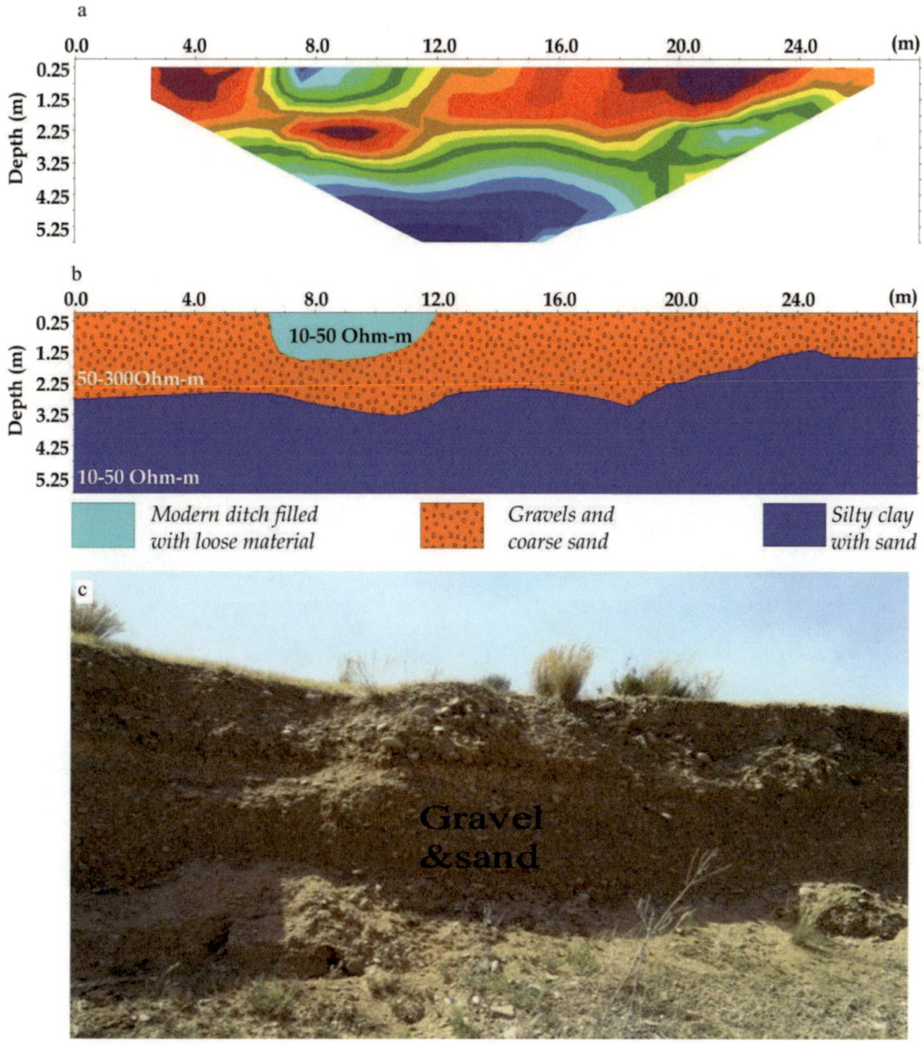

Figure 6. a) 2D resistivity section that was extracted from the 3-D inversion model. b) Geologic interpretation of the different geoelectrical units. c) In situ stratigraphic layers showing the gravel and sand layer.

14:40, the resistivity signature decreased to more than 20%. Concurrently, after the 4th time phase, the ERT measurements also show a resistivity decrease that was formulated within a circular region at the middle of the section (X=13-15m, Z>2.5m). The resistivity decrease initiates from a value of about 6% at 12:03 hours by the end of Phase T4 and reaches almost 16-18% at the end of the experiment. A similar signature was registered along the profile L8, where within four hours of monitoring (10:44pm – 14:52) the resistivity was decreased more than 20% (Fig. 7). The area that is occupied by the drop of resistivity lies within the horizontal distance X=8 – 11m and depth more than one meter. A similar signature also appears at the beginning of the Line 8 (X=4-6.5m) (Fig. 7).

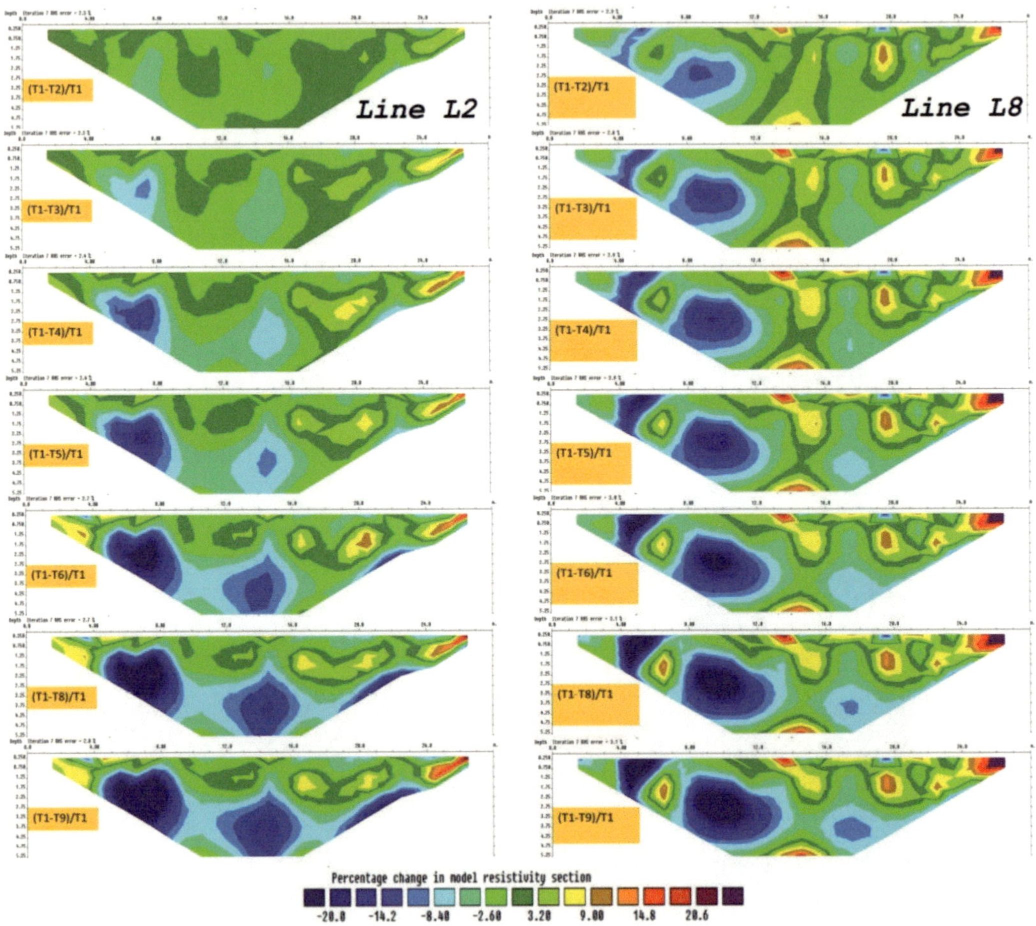

Figure 7. Sections of the % subsurface resistivity change with time for phases T2 to T9 with respect to the reference phase T1 for Lines L2 and L8. Phase T7 was excluded from analysis due to erroneous data.

Afterwards the ERT images for Lines L2 and L8 showing the relative percentage change of resistivity of the later time phases (T2 to T9) with respect to the first phase (T1) were overlaid on the plan view of the site. The resistivity decrease along the 2-D sections reflects the water flow within the subsurface. This was projected on the ground surface in order to outline the preferred water flow paths within subsurface, at a depth of more than one meter. The water due to the leakage of the pipe shows a selective flow towards the east along three main paths (Fig. 8).

The resulting GPR slices for 1.0m depth are presented in Figure 10. GRID0 illustrates the subsoil before the water leakage starts. Two strong anomalies appeared at Y=20m, X=6-14m and Y=5,

X=7-15m. These anomalies also appeared in all the phases after the leakage started. The difference is that both of the anomalies extend towards the location of the pipe that ejects the water. As well, the anomalies tend to be more linear in comparison with the dry soil, especially for GRID7 (Fig.9). It is clear that those displacements and changes of the anomalies' location and shape are due to the water leakage.

Figure 8. Outline of the footprint of water flow on the ground surface based on the results of the ERT monitoring experiment.

In Figure 10 and 11, the percentage difference of GRID0 and GRID1 respectively are presented. For the first case (Fig. 10) the differences in instantaneous amplitude don't exceed 50% and seems to be the same for all the phases. There is not a significant fluctuation of the amplitudes among the different phases to provide information related to the water paths propagation in time. The second case (Fig. 11) showed no changes in amplitude values. The water flow may have caused changes in the existing anomalies' shape and location but it didn't affect the amplitude values.

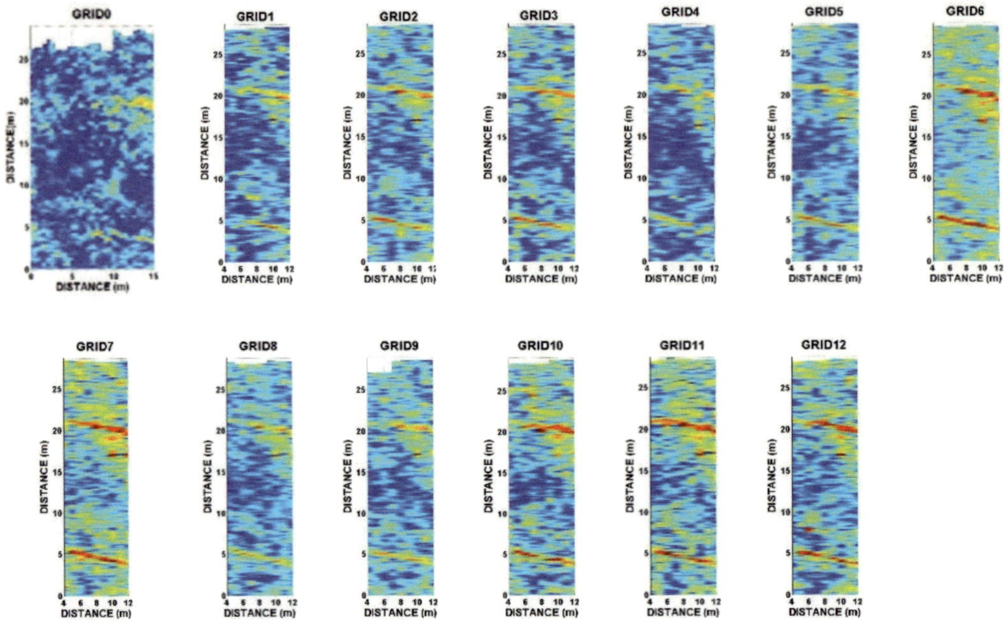

Figure 9. Resulting GPR slices during each phase for 1.0m depth.

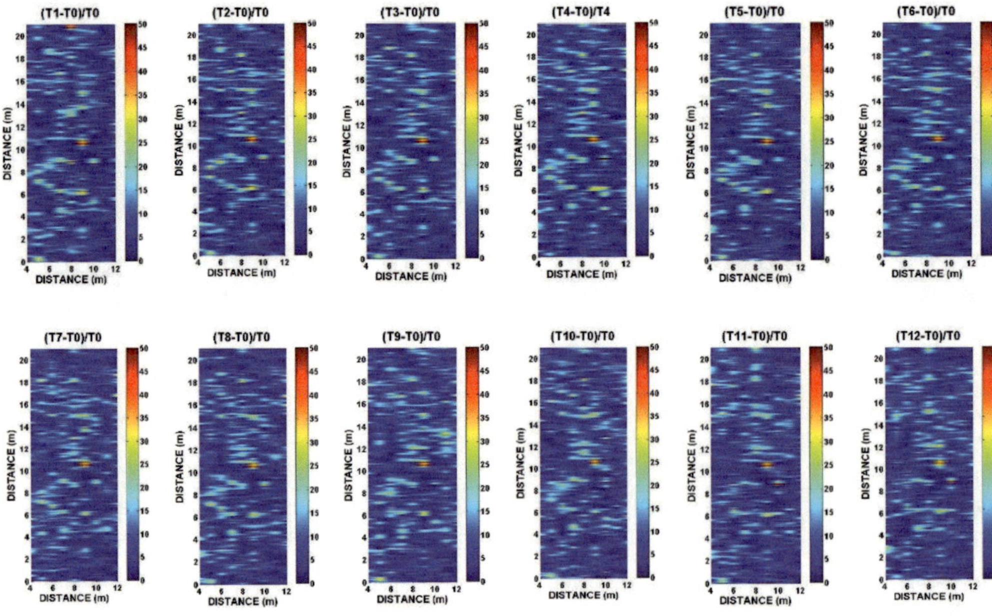

Figure 10. Percentage differences in respect to the first phase of dry soil.

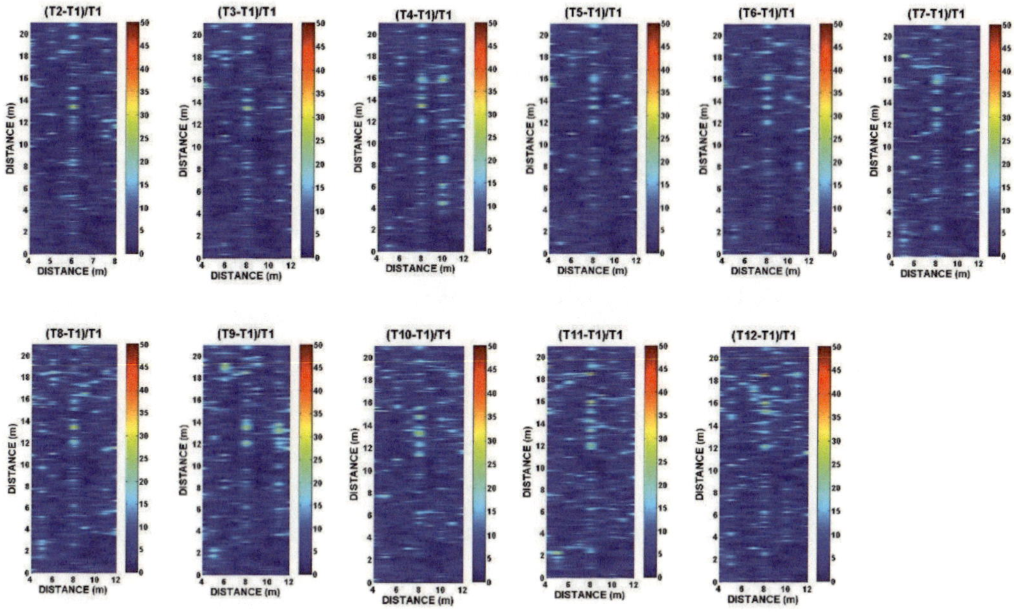

Figure 11. Percentage differences in respect to the second phase, during which the water leakage started.

7. Conclusions

The results of the controlled experiments demonstrated the effectiveness of ERT in locating and monitoring the water flow along preferred water paths. In contrast to ERT, GPR exhibited inferior results. The changes on the existing reflectors could not give a distinct signature in order to describe the water paths. This does not necessarily mean that GPR is not an appropriate method for monitoring water leakages. The poor results can be attributed to the fact that the water paths appear at a depth of more than 1.0 m where the conductivity is very high due to the presence of clay resulting increased absorption for the GPR signal.

Acknowledgements

The results reported here are based on findings of the Cyprus Research Promotion Foundation project "ΑΕΙΦΟΡΙΑ/ΦΥΣΗ/0311(ΒΙΕ)/21": Integrated use of space, geophysical and hyperspectral technologies intended for monitoring water leakages in water supply networks in Cyprus. The project is co-funded by the Republic of Cyprus and the European Regional Development Funds of the EU.

Author details

Meropi Manataki[1], Nikos Papadopoulos[1], Apostolos Sarris[1], Athos Agapiou[2], Kyriacos Themistocleous[2] and Diofantos G. Hadjimitsis[2*]

*Address all correspondence to: d.hadjimitsis@cut.ac.cy

1 Laboratory of Geophysical-Satellite Remote Sensing & Archaeo-environment (GeoSat ReseArch), Institute for Mediterranean Studies (I.M.S), Foundation for Research & Technology, Hellas (F.O.R.T.H.), Nik. Foka 130, Rethymno 74100, Crete, Greece

2 Department of Civil Engineering and Geomatics, Faculty of Engineering and Technology, Cyprus University of Technology, Cyprus

References

[1] Annan, A.P.. GPR- History, Trends and Future Developments. *Subsurface Sensing Techonology Applied*, 2002. 3: pp. 253-270.

[2] Aspinall, A and Lynam, J.T.. An induced polarization instrument for the detection of near-surface features. *Propezioni Archaeologica*, 1970. 5: pp. 67-75.

[3] Atzemoglou, A., Tsourlos, P., and Pavlides, S.. Investigation of the tectonic structure of the NW part of the Amynteon Basin (NW Greece) by means of a vertical electrical sounding (VES) survey. *Journal of the Balkan Geophysical Society*, 2003. 6: pp. 188-201.

[4] Benson, A.K. . Applications of ground penetrating radar in assessing some geological hazards: examples of groundwater contamination, faults, cavities.. *Journal of Applied Geophysics*, 1995. 33: pp. 177-193.

[5] Bimpas, M., Amditis, A., and Uzunoglu, N.. Detection of water leaks in supply pipes using continuous wave sensor operating at 2.45GHz. *Journal of Applied Geophysics*, 2010. 70: pp. 226-236.

[6] Cassidy, N.J. . Chapter 2 - Electrical and Magnetic Properties of Rocks, Soils and Fluids, in:Harry M. Jol (eds.). *Ground penetrating Radar Theory and Applications*. Amsterdam: Elsevier. 2009a. 41-72.

[7] Dahlin, T. and Owen, R.. Geophysical investigations of alluvial aquifiers in Zimbabwe, in: *Proceedings of the IV Meeting of the Environmental and Engineering Geophysical Society (European Section)*. Barcelona, Spain. 1998. 151-154.

[8] Dahlin, T., Johansson, S., and Landin, O.. Resistivity Surveying for Planning of Infrastructure., in: Environment and Engineering Geophycial Society (eds.). *Symposium on the Application of Geophyscis to Entigeering and Engineering Problems*. 1994. 509-528.

[9] Davis, J.L. and Annan, A.P.. Ground- Penetrating Radar for High-resolution Mapping of Soil and Rock Stratigraphy. *Geophysical Prospective*, 1989. 37: pp. 531-551.

[10] Doolittle, J.A., Jenkinson, B., Hopkins, D., Ulmer, M. and Tuttle, W. . Hydropedological investigations with ground penetrating radar (GPR): Estimating water-table depths and local ground-water flow pattern in areas of coarse-textured soils. *Geoderma*, 2006. 131: pp. 317-329.

[11] El-Said, M.A.H.. Geophysical Prospection of Underground Water in the Desert by Means of Electromagnetic Interference Fringes. *Proc. IRE*, 1956. 44: pp. 24-30.

[12] Grasmueck, M.. 3-D ground-penetrating radar applied to fracture imaging in gneiss. *Geophysics*, 1996. 61: pp. 1050-1064.

[13] Hunaidi, O., Chu, W., Wang, A., and Guan, W. . Leak Detection Methods for Plastic Water Distribution Pipes, American Water Works Association. 1999.

[14] Jordana, J., Gasulla, M., and Pallas-Areny, R.. Electrical resistance tomography to detect leaks from buried pipes. *Measuring Science Technology*, 2001. 12: pp. 1061.

[15] Karaoulis, M.C., Kim, J.H. and Tsourlos, P.I. . 4D active time constrained resistivity inversion. *Journal of Applied Geophysics*, 2011. 73: pp. 25-34.

[16] Kim, J.H., Yi, M.J., Park, S.G. and Kim, J.G.. 4-D inversion of DC resistivity monitoring data acquired over a dynamically changing earth model. *Journal of Applied Geophysics*, 2009. 68: pp. 522-532.

[17] Legault, J.M., Carriere, D., and Petrie, L. . Synthetic model testing and distributed acquisition dc resistivity results over an unconformity uranium target from the Athabasca Basin, northern Saskatchewan.. *The Leading Edge*, 2008. 27: pp. 46-51.

[18] Loke, M. and Barker, R.D.. Rapid least-squares inversion of apparent resistivity pseudosections by a quasi-Newton method. *Geophysical Prospective*, 1996. 44: pp. 131-152.

[19] Loke, M., Acworth, I. and Dahlin, T.. A comparison of smooth and blocky inversion methods in 2D electrical imagining surveys.. *Exploring Geophysics*, 2003. 34: pp. 182-187.

[20] Mares, S.. *Introduction to Applied Geophysics*. Sppringer Science and Business Media. 1984.

[21] Panissod, C., Dabas, M., Hesse, A., Jolivet, A., Tabbagh, J., and Tabbagh, A. . Recent developments in shallow-depth electrical and electrostatic prospecting using mobile arrays. *Geophysics*, 1998. 63: pp. 1542-1550.

[22] Papadopoulos, N. Sarris, A., Yi, M. and Kim, J. . Urban archaeological investigations using surface 3D Ground Penetrating Radar and Electrical Resistivity Tomography Methods. *Exploring Geophysics*, 2009. 40: pp. 56-68.

[23] Papadopoulos, N.G., Tsourlos, P., Tsokas, G.N. and Sarris, A.. Two-dimensional and three-dimensional resistivity imaging in archaeological site investigation.. *Archaeological Prospective*, 2006. 13: pp. 163-181.

[24] Papadopoulos, N.G., Tsourlos, P., Papazachos, C., Tsokas, G.N., Sarris, A. and Kim, J.H.. An algorithm for fast 3D inversion of surface electrical resistivity tomography data:application on imaging buried antiquities. *Geophysical Prospective*, 2011. 59: pp. 557-575.

[25] Powers, M.H., and Olhoeft, G.R. . Moeling the GPR Response of Leaking, Buried Pipes. *EEGS*, 1996. : pp. 525-534.

[26] Ramirez, A., Daily, W., Binley, A. and LaBrecque, D. . Tank Leak Detection Using Electrical Resistance Methods, in:Environment and Engineering Geophysical Society (eds.). *Symposium on the Application of Geophysics to Engineering and Environmental Problems*. Environment and Engineering Geophysical Society. 1996. 763-772.

[27] Spanoudakis, S.N. and Vafidis, A. . GPR-PRO: A MATLAB module for GPR data processing, in:*Proceedings of the 2010 13th International Conference on ground Penetrating Radar (GPR)*. 2010. 1-5.

[28] Spanoudakis, N.S., Manataki, M., Niniou-Kindeli, V. and Vafidis, A.P. . GPR Imaging at Aptera Archaeological Site, in:*Proceedings of the 6th Congress of Balkan Geophysical Society*. 2011.

[29] Stampolidis, A., Soupios, P., Vallianatos, F. and Tsokas, G.N.. Detection of leaks in Buried plastic water distribution pipes in urban places-a case study, in:*Proceedings of the 2nd International Workshop on Advanced Ground Penetrating Radar*. 2003. 120-124.

[30] Theune, U., Rokosh, D., Sacchi, M. and Schmitt, D.. Mapping fractures with GPR: A case study from Turtle Mountain.. *Geophysics*, 2006. 71: pp. B139-B150.

[31] Tzanis, A.. Matgpr: a freeware matlab package for the analysis of common-offset GPR data.. *Geophysical Research Abstracts*,

[32] Vaughan, C. . Ground-peentrating radar surveys used in archaeological investigations. *Geophysics*, 1986. 51: pp. 595-604.

[33] Weymouth, J.W. and Huggins, R.. Geophysical Surveying of Archaeological Sites, in:Rapp, G and Gifford, J. (eds.). *Archaeological Geology*. 1985. 191-236.

Chapter 4

The Use of GIS for Supporting the Experimental Representation of the Selected Supply Network in Pafos Municipality — 'The Hydrogis Lab'

Kyriacos Themistocleous, Ploutarchos Evlogimenos,
Athos Agapiou and Diofantos G. Hadjimitsis

Additional information is available at the end of the chapter

http://dx.doi.org/10.5772/59533

1. Introduction

The HydroGIS Lab project (http://cyprusremotesensing.com/hydrogis/) aims to satisfy the dire need for authorities to solve the extremely serious problem of water supply as a result of continued water shortage. The chronic problems of water losses in the water supply network pipes, the uncontrolled and non-optimum operation of pumping stations, the incorrect design of the networks because of various interventions such as town planning and erroneous mapping of existing networks, are some of the most important problems which need to be resolved in order to optimize the performance of the networks and, consequently, save water and energy. The innovative aspect of the Project is that state-of-the-art technologies are combined for the mapping of water networks through the Global Positioning System (GPS), Radar Scanners and Satellite Remote Sensing (SRS). The data is entered into a Geographic Information System (GIS), with the aim of developing a digital imprint and the mapping of the network. The study of an existing selected water supply network and the collection of the necessary information by the Cyprus University of Technology and the Water Development for the network and its systems will lead to an imaging of water pipes and their design using a radar scanner and special software and GIS. Satellite images, which have been transformed into the local reference system using specialized software, will be coupled with the digital imaging of the existing maps. This information will then be organized in such a way allowing the development of maps and their analysis at different levels (e.g. water supply network, buildings, water supply elements, geographic information, features of the network's systems, etc.).

GIS has been used to develop a comprehensive management system of the information, including text, cartographic and quantitative data. Maps from Paphos Municipality depicting existing buildings, streets and other infrastructure were used in creating the digital map. The ground penetrating radar (GPR) provided imaging of all required information regarding the existing water supply pipes, including altitudes, grading, location of other services, diameters, etc. Satellite images of the region of Paphos Municipality will be used to retrieve the present conditions, such as current construction development and, therefore, the immediate water supply needs in the municipality.

2. Methodology

A representative section of the water supply network in the Pafos Municipality area, including pumping stations, reservoirs, valves, and flowmeters, were recorded and the selected network was digitized. The MALA GPR was used to conduct a survey of the existing water network. After the GPR survey was concluded, the GPR data were analyzed and the water pipes were identified, according to their diameter, material and depth. Following, each survey point was inserted into a GIS system, including all the data acquired. The network was then connected from each survey point.

The power of a GIS comes from the ability to correlate different information in a spatial context and to reach a conclusion about this relationship. All the retrieved information that were available through the acquired software regarding surveying, analysis, management, processing, were inserted in the GIS database. The GIS system is flexible, allowing the addition of any layer considered necessary for the better management of the Project information. The final structure and content of the digital imprint will depend on the results of the initial study of the networks in Paphos Municipality, in order to proceed with an accurate modeling of the representative network.

A methodology will be created for detecting and adapting critical parameters of the mathematical models (e.g. surface roughness coefficient of pipes etc.), based on actual experimental measurements of the flow field in corresponding hydraulic models. As well, detailed experimental verification of the computational results and the development of knowledge for self-regulation of pumps and valves will be examined in order to satisfy the supply needs in conjunction with the uniform pressure distribution in the water supply network. In this way, guidelines will be developed which could form the basis for developing a user-friendly network simulation software, able to make up for the lack of communication between the different sciences involved in developing, designing and managing water supply networks.

A framework was created to develop specifications and relevant computational tools for the optimum analysis of supply networks, both in the design and the operation phase. An optimization framework was defined, including the identification of the problem (determination of control variables and performance criteria) and its solution, using multicriteria evolution algorithms. This methodology will be utilized in the context of an information system for the management of networks, which will include a database, a GIS and a compu-

http://dx.doi.org/10.5772/59533

tational code. In order to implement the computational code, existing tools and models will be used or new ones will be developed, so as to ensure compatibility with the proposed methodological framework.

The operation of the experimental model is studied in detail, with the support of specialized instruments and systems for the continuous recording of the variables (i.e. flow capacity, pressure, power etc) at different points in the network. The experimental results will be utilized for verifying the theoretical models as well as for developing a framework for the identification and optimum adaptation of critical parameters of the simulation models (e.g. surface roughness coefficients), based on the measurements in the corresponding hydraulic model. Furthermore, specialized analyses will be conducted in typical locations of the experimental model, which may prove to be problematic (e.g. cross-sections downstream and upstream of the pumping stations) in order to study flow phenomena (e.g. local losses at operational phases) and propose improvements.

3. Ground-penetrating radar (GPR)

Current research (Eyuboglu et al, 2004; Demirci et al., 2012; Gamba and Lossani, 2000) states there is a trend toward development of more and more sophisticated systems including the ground penetrating radar (GPR) technique, which is safe for use in urban environments and also protects the geological, environmental and archaeological integrity of subsurface settings (Gamba and Lossani, 2000). GRP is a geophysical method that uses radar pulses to image the subsurface. It is a nondestructive method that uses electromagnetic radiation in the microwave band of the radio spectrum, and detects the reflected signals from subsurface structures (Dong et al., 2012). GPR interpretation of this reflected energy may yield information on subsurface structural variation and condition of the pipeline, as indicated in figure 1.

GPR is also one of the most used tools in the field of the water monitoring and management especially in the fields of the drainage pipes detection and characterization, water leaks in pipe detection and determination of the time-behavior of water content in the soil (Allerd et al., 2008; Crocco et al.,2010; Lambot et al.,2008; Hadjimitsis et al., 2013). GPR can accurately pinpoint buried pipeline leaks without digging. The leaking substances can be identified at the source by the radar via the changes in the surrounding soil's electrical parameters. The GPR is able to generate 3D underground images of pipes, power, sewage and water mains. As well, it can identify leaks in buried water pipes either by detecting underground voids created by the leaking water as it erodes the material around the pipe, or by detecting anomalous change in the properties of the material around pipes due to water saturation. Unlike acoustic methods, application of ground penetrating radar for leak detection is independent of the pipe type (e.g., metal or plastic) (Eyuboglu et al, 2004). The principles involved are similar to reflection seismology, except that electromagnetic energy is used instead of acoustic energy, and reflections appear at boundaries with different dielectric constants instead of acoustic impedances.

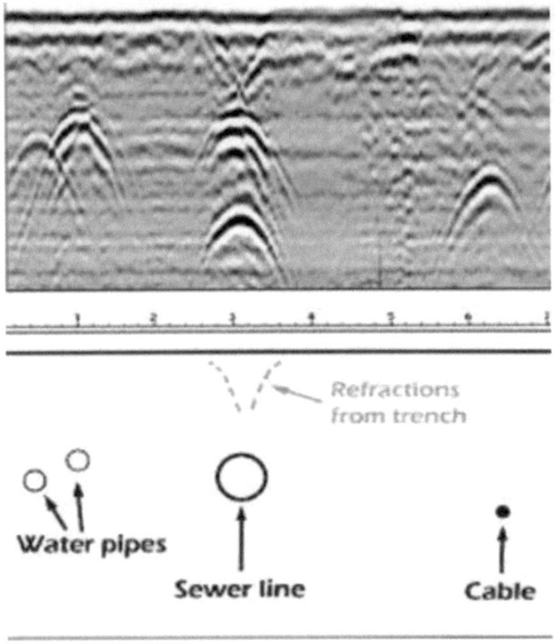

Figure 1. Reflected signals from subsurface structures

4. Survey and GIS

Paphos Municipality is one of the municipalities in Cyprus facing very serious problems with the water supply network, due to the age of the network, rapid town development and the expansion of the network in an unorganized manner. One of the aims of the HYDROGIS project is the study of an existing selected water supply network in the Municipality in an effort to understand the 'logic' of the network and select a representative part of it that will include almost all the components (i.e. pumping station, main piping, valves, flowmeters etc) present in the networks of the Municipality in order to conduct a comprehensive study. Therefore, the mapping and development of a database using GPS and GIS technologies, as featured in figure 2, has the aim of developing a digital imprint of a representative water network in Paphos Municipality and the creation of scale modeling for detailed experimental studies.

A representative section of the water supply network in the Pafos Municipality area, all the features and systems which compose it, including pumping stations, reservoirs, valves, and flowmeters have been recorded and the selected network was digitized (Figure 3). A survey through the use of a latest technology radar scanner has been conducted. The specific radar scanner is a MALA ground penetrated radar designed for urban areas and background noise. The shielded antenna used consists of both transmitter and receiver antenna elements in a single housing. After the GPR survey was concluded, the GPR data were analyzed using the GroundVision 2 and RadExplorer software, where the pipes were identified, according to their diameter, material and depth. Following, each survey point was inserted into a GIS system, including all the data acquired. The network was then connected from each survey point, as indicated in Figure 4.

Figure 2. Geo-radar Scanning of the Paphos water utility network

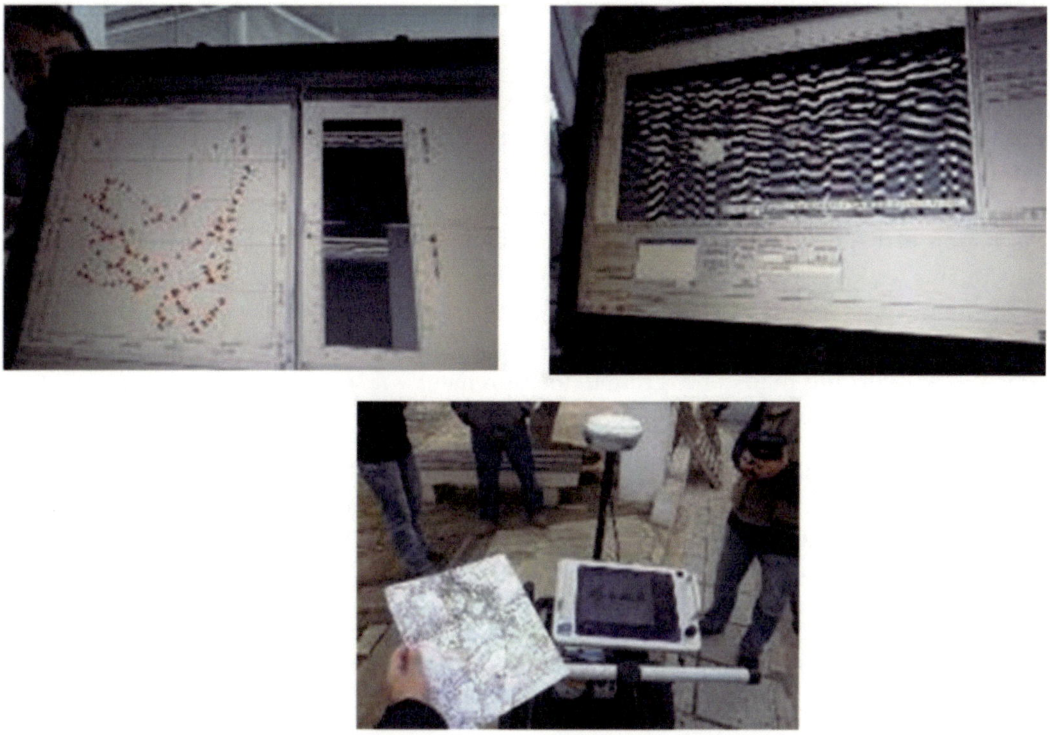

Figure 3. study, mapping and development of a digital imprint of an existing selected water supply network

Figure 4. The water network following the GPR survey

GIS was used to develop a comprehensive management system of the information, including cartographic and quantitative data, as well as text. Maps from Paphos Municipality depicting existing buildings, streets, etc. were used in creating the digital map. The use of the radar scanner permitted the imaging of all required information regarding the existing water supply pipes, including altitudes, grading, location of other services, diameters, etc. Satellite data of the region of Paphos Municipality, including Quickbird, IKONOS, Landsat TM or SPOT satellite images, will be used to retrieve the present conditions. The capturing of satellite data at any moment in time using remote sensing techniques can provide the current construction development and, therefore, the immediate water supply needs in the municipality. In conjunction with the maps and ground measurements, it will be possible to develop an accurate digital imprint.

The acquired information will be organised in a GIS system, thereby allowing the creation of maps and their analysis at different "layers", including the water supply network, buildings, water supply elements, geographic information, features of the network's systems, etc. The basic structure of the layers is described as follows:

The first layer depicts the connections of the pipes to the buildings, including information relating to serviced flowmeters, the main input pipe, the relevant dimensions of the pipes, etc.

The second layer depicts the pipes of the network. It includes information on the geometrical elements of the excavation where the pipe is located, features of the wells, the length of the pipe, altitudes of the pipes determining their grading, etc.

The third layer will depict the pumping stations of the selected network and will include, in text form, all the features of the pumping station such as supplies, pressures, features of the systems comprising the pumping station such as pumps, valves, flowmeters and relief systems, etc.

The fourth layer will depict the reservoirs together will all the junctions/joints of the pipes and will include a description of all the features of the reservoirs.

The fifth layer will be the type of material of the pipe network, the diameter, the depth of the network, and the installation year. Moreover, attribute tables regarding water consumption, flow, water leakages records etc. will be included.

The sixth layer will be used as a background layer from cadastral maps, elevation, etc.

The seventh layer will be the results of the field study of the radar measurements that will be surveyed by GPS technology.

Finally all the retrieved information that are available through the acquired software regarding surveying, analysis, management, processing, will be inserted in the GIS database (Figure 5). The GIS system will be flexible, allowing the addition of any layer considered necessary for the better management of the Project information. It must be noted that depending on the results of the initial study of the networks in Paphos Municipality, the final structure of the digital imprint as well as necessary information to be contained therein will be decided, in order to proceed with an accurate modeling of the representative network.

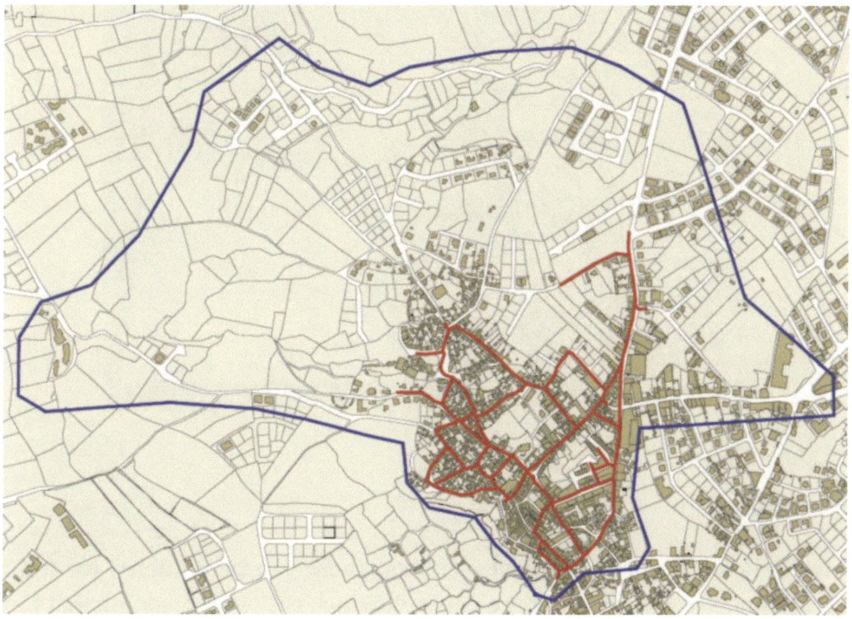

Figure 5. Area of interest, including the GPR water pipe network survey

5. Distribution network

The main consideration in the design and evaluation of a water distribution system is the ability of the network to supply consumers with water of adequate quantity and quality, for the expected loading conditions during its lifetime. A reliable water distribution system could perform within specified limits in both normal and abnormal conditions, such as breaks in pipes, mechanical failure of pipes, valves, and control systems, power outages, malfunction of storage facilities and inaccurate demand projections (T. Akdoğan, 2005). Several analytical methods exist, that lead to the calculation of surrogate – based measures of reliability, which take into account the reliability of every mechanical part of the network and statistical distribution of nodal demands and loading scenarios (figure 6). Since these methods require full knowledge of the network's layout and function, stochastic simulation methods for the estimation of network reliability are often preferable, as they can provide flexibility in the types of network elements that can be included in analysis (Wagner et al., 1988), resulting in a more realistic estimation of the network reliability. The water supply network is evaluated using different user defined scenarios or during extended period simulations (Goulter et al., 2000).

Figure 6. Typical 3D drawing of a distribution network

The local network of Moutallos district has been chosen to be modeled for the following reasons: (1) It is an old network. (2) It is a network which is closed, meaning that there is only inflow and consumption, therefore one can easily estimate the water loss due to network leakage. (3) It is limited to a small area, so that it can be modeled accurately.

http://dx.doi.org/10.5772/59533

6. Laboratory model

The laboratory modelling was implemented in the laboratory area located at the Industrial area of Agios Silas in Limassol in Cyprus. The water supply of Moutallos comes from a storage tank which is supplied with water from a pumping station via two pumps working in parallel with flow capacities 160 and 180 m3/hour respectively. The water supply to the laboratory constant head tank was implemented by a pump of maximum capacity 350 m3/hour, which may vary using an AC driver to control the rpm of the pump motor, thus varying the flow rate delivered.

The water supply network that was designed included a constant head tank with volume of 25 m3, a storage tank of 1000 m3 located in the basement area, pipes of 3, 6 and 8 inch diameters, a pump with motor and decompression valve, analog valves for flow control and local loss introduction, flow meters, piezometers, thermometer, and hand valves for manual control of the network (Figures 7-15). All these devices, except the hand valves, can be controlled and recorded electronically. To simulate the reservoir that supplies water to the Municipality of Pafos (Moutallos area) with gravity, a constant head tank was constructed that can supply the pipe network with 9 m pressure head. Thus, keeping the reservoir head constant via a return overflow to the storage reservoir, experiments and computer simulations were conducted to manipulate the flow using the different valves according to the desired scenario.

Figure 7. Photo of a part of the pipe network as it has been constructed

Figure 8. Pipe network during construction

Figure 9. Analog valve (left) and pressure relief valve (right)

http://dx.doi.org/10.5772/59533

Figure 10. Digital flowmeter

Figure 11. Constant head tank

Figure 12. Constant head tank, connection to the experimental network

Figure 13. Overview of the pipe network

Figure 14. Overview of special network items such as bends, junctions, etc.

Figure 15. Long straight pipe segments where the friction losses as measured.

Two separate flow networks were constructed. The first was supplied from the constant head tank with hydrostatic pressure of nine meters. The second was connected to a 75 KW pump which is controlled by an AC Driver so that variable flow conditions can be applied to the network. In the second case, the consumption of electricity as a function of the load applied to the network was able to be recorded. The two flow networks have a length of 100 m each and are made of 3 and 6 inch pipes respectively. They can be used separately or they can be combined to one system of pipes with two different diameters. In both networks measurements can be made at several points, such as inflow discharge using flowmeters, pressure measurement along a straight pipe, at bends, at a change of diameter, etc for evaluating the friction (linear) and local energy loss of the pipe network that can be used in Moutallos network modeling via Epanet.

In the experimental pipe network, multiple sensors have been installed such as electronic manometers (9 pieces), electronic thermometers (8 pieces), magnetic-inductive flowmeters (2 pieces) torque meter (1 piece), ammeter (1 piece). They were all connected to a data acquisition system on a computer where the transmitting signal can be sampled in real time, in order to create a computer model of the pipe network.

In addition to the water supply to the pipe network from the constant head tank which is controlled by valves, the network was also supplied with the desired flow rate using the variable speed AC DRIVER motor of the pump, thus increasing the pipe network to the desired pressure as well. The energy losses of the whole network can be estimated using the power consumption data of the pump, which can be recorded using the data acquisition system installed in the facility.

The energy loss along each pipe (friction losses) and at special items as bends, junctions, pipe expansion or contraction, valve, etc. (Figure 14 and 15) can be estimated from the pressure difference measured at the two ends of the pipe or the special part of interest as a function of the flowrate. The piezometers have been installed at least 10 to 15 pipe diameters away from changes in the flow direction, so that the flow at the pipe section where the pressure is measured is uniform and not affected by secondary flow.

6.1. Pipe aging

One of the problems in the computer modelling of the pipe network is the aging of the pipes in Moutallos area. The age of the simulated network played an important role in the calculation, since the accumulation of salts may reduce the pipe diameter and also increase the roughness. Aging of the network can be obtained by keeping the pipe diameters the same while increasing the roughness in order to obtain the same friction loss as the actual aged pipe. Pipe aging can be measured through estimation of the discharge using some direct (volumetric) or indirect methodology in approximately 10 m long pipe segments for each pipe diameter used in Moutallos, along with the corresponding energy (pressure) loss. Thus, actual friction loss can be estimated by in-situ measurements along a straight pipeline, by recording the flow rate and the pressure loss simultaneously over a distance where there is no water supply to nearby houses during data acquisition. This is usually performed to as many pipes as the ages of which appear in the network. For example, if 6 inch pipes were installed in the network 35, 20 and 8 years ago, then three measurements should performed, for the pipe of every age, so that estimations of the pipe roughness variation over the years (35, 20 and 8).

6.2. Flowrate estimation

The range of flow rates used in each pipe to be modelled can be deduced from the real flow rates that may occur in the network to be modelled. Thus, from the computer model of the Moutallos water distribution network where the water demand at each node has been estimated properly, the instantaneous minimum as well as the maximum demands for the area under consideration, thus estimating the regime of the discharge magnitude that is possible to occur in each different pipe diameter. To date, the experiments have been delayed due to the electrical connections of the sensors that have been installed in the pipe network with the data acquisition system, as well as the implementation of the proper data acquisition module in the LabView software environment which has been chosen to be used. The experimental pipe network has been tested properly and it is operational.

7. Methodology

The pipe network of Moutallos is modelled and solved in the commercially used EPANET software platform (Figure 16). The method used in EPANET to solve the flow continuity and headloss equations is called "Gradient Method" (Todini and Pilati, 1987 and Salgado et al., 1988). The method is thoroughly described in Epanet User's Manual (Rossman, 2000). Epanet can perform single or extended period analysis modelling of the hydraulic and water quality behavior of pressurized pipe networks that consist of pipes, nodes, valves, and storage tanks or reservoirs. EPANET's Windows user interface provides a visual network editor that simplifies the process of building piping network models and editing their properties and data. Color-coded network maps, data tables, and contour plots and other reporting and visualization tools assist in interpreting the results of a network analysis.

The water distribution network is modelled in Epanet and stochastically analyzed using Epanet Programmer's Toolkit. EPANET and its Programmer's Toolkit have been developed by the Water Supply and Water Resources Division of the U.S. Environmental Protection Agency's National Risk Management Research Laboratory, and are both freely available in the internet.

The process to built the network using the Epanet user interface is as follows:

The existing network layout plan was loaded as background map.

Tanks and nodes of the network were placed. Nodes are located in the junctions, in dead-ends of the network and where pipe diameter or material changes. Required information for every node is its elevation and base demand, and additionally initial water level for tanks.

Pipes connecting nodes based on the network layout were added. In Property Editor, the length, diameter and roughness for every pipe were entered.

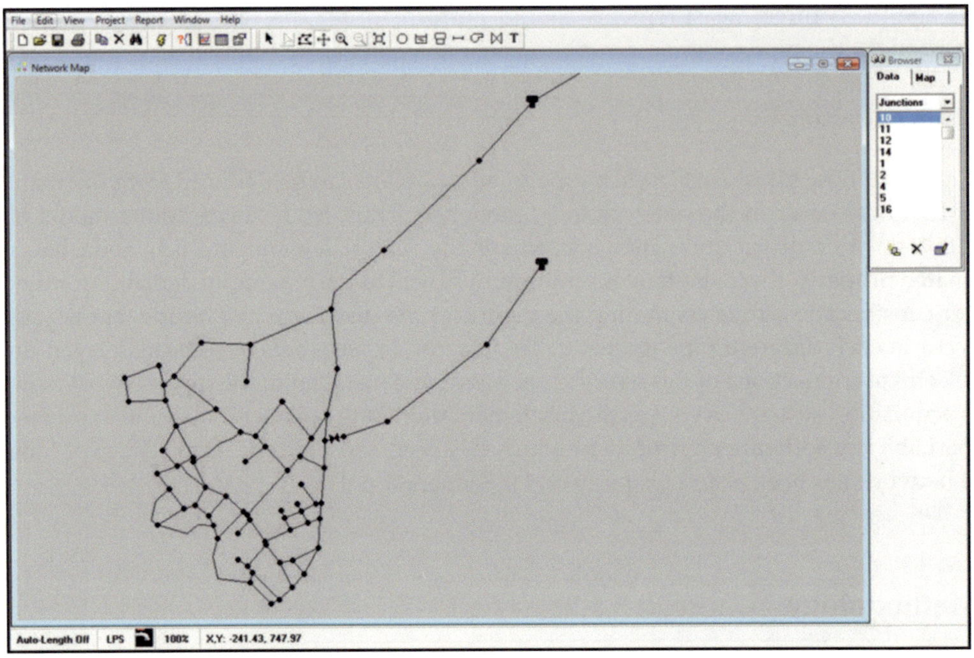

Figure 16. Model of pipe network of Moutallos in Epanet environment

The Darcy – Weisbach equation was selected as default formula for linear headloss calculations.

Project's data were exported to a text file (.inp), so that EPANET can run as a console application.

8. Results

The HydroGIS laboratory provides a model that is used to simulate various water supply networks. A methodology was created for detecting and adapting critical parameters of the mathematical models (e.g. surface roughness coefficient of pipes etc), based on actual experimental measurements of the flow field in corresponding hydraulic models. Detailed experimental verification of the computational results and the development of knowledge for self-regulation of pumps and valves were examined to satisfy the supply needs in conjunction with the uniform pressure distribution in the water supply network. In this way, guidelines will be developed which could form the basis for developing a user-friendly network simulation software, able to make up for the lack of communication between the different sciences involved in developing/designing and managing water supply networks.

The aim of the project is to utilize scanning and 3D spatial representation to develop a GIS platform for managing all information required, including maps, satellite imaging, 3D scans, network system components and others. A methodology is being developed regarding the critical parameters required for hydraulic models, and is being tested in the HydroGIS laboratory. Based on the digital imprint of the selected network, construction drawings of the network were modeled and included all the necessary information of the piping and locations of all piping components, fittings and configurations. By using the HydroGIS laboratory, the assembled network with all its necessary components process of "logical" construction of the piping were modeled in real life in order to identify the components that need to be replaced in order to reduce the water leakages in the network. Since the pipe network has been modeled, over 100 computer simulations of the network have been performed. From these simulations, node demands were estimated from yearly consumption. The network reliability is reduced for increment in hourly demand greater than 20-30% and CV 15-20%. However, more measurements are needed for more accurate design. As well, laboratory data regarding local or friction losses for different types of pipes (if available), can be applied in future pipe network modelling.

Acknowledgements

The results reported here are based on findings of the Cyprus Research Promotion Foundation project "ΑΝΑΒΑΘΜΙΣΗ/0609/34": Upgrade of the hydraulics laboratory for the modeling of water supply networks & design and operation optimization study. The project is co-funded by the Republic of Cyprus and the European Regional Development Funds of the EU.

Author details

Kyriacos Themistocleous, Ploutarchos Evlogimenos, Athos Agapiou and Diofantos G. Hadjimitsis*

*Address all correspondence to: d.hadjimitsis@cut.ac.cy

Department of Civil Engineering and Geomatics, Faculty of Engineering and Technology, Cyprus University of Technology, Cyprus

References

[1] Akdogan, T.. Design of water distribution system by optimization using reliability considerations [Dissertation]. Middle East Technical University.

[2] Allred, B.J., Daniels, J.J. and Ehsani, M.R.. *Handbook of Agricultural Geophysics*. CRC Press. 2008.

[3] Crocco, L., Soldovieri, F., Millington, T. and Cassidy, L.. Bistatic tomographic GPR imaging for incipient pipeline leakage evaluation. *Progress in Electromagnetics Research*, 2010. PIER: pp. 307-321.

[4] Demirci, S., Yigit, E. Eskidemir, I.H. and Ozdemir, C.. Ground penetrating radar imaging of water leaks from buried pipes based on back-projection method. *NDT & E Internation*, 2012. 47: pp. 35-42.

[5] Dong, L., Carnalla, S.and Shinozuka, M.. GPR survey for pipe leakage detection: experimental and analytical study, in:Gyekenyesi, A.L. (eds.). *Proceedings of the SPIE 8347*. SPIe. 2012. article id 83470F.

[6] Eyuboglu, S. Mahdi, H. and Al-Shukr,i H.. Detection of water leaks using ground penetrating radar [Internet]. 2004. Available from: http://www.dot.state.fl.us/statematerialsoffice/geotechnical/materials/eyuboglu-mahdi-alshukri.pdf

[7] Gamba, P. and Lossani, S.. Neural detection of pipe signatures in ground penetrating radar images. *IEEE Transactions on Geosciences and Remote Sensing*, 2000. 38(2)

[8] Goulter, I., Walski, T.M., Mays, L. Sekarya, A.B.A., Bouchart, R. and Tung, Y.K.. *Reliability analysis for Design-Water distribution systems handbook*. 2000.

[9] Hadjimitsis, D.G., Agapiou, A., Themistocleous, K., Toulios, G., Perdikou, S. Toulios, L. and Clayton, C.. *Detection of water pipes and leakages in rural water supply networks using remote sensing techniques*. 2013. doi: isbn 978-953-51-1152-8

[10] Lambot, S, Slob, E.C., Chavarro, D., Lubszynski, M. and Vereecken, H. . Measuring soil surface water content in irrigated areas of southern Tunisia using full wave inversion of proximal GPR data. *Near Surface Geophysics*, 2008. 16: pp. 403-410.

[11] Rossman, L.A.. *EPANET Programmer's toolkit manual*. . Cincinnati, Ohio: Watter Supply nad Water Resources Division, National Risk Management Research Laboratory, US Environmental Protection Agency. 2000.

[12] Slagado, R., Todini, E., and O'Connel, P.E.. Comparison of the gradient method with some traditional methods for the analysis of water supply distribution networks, in:*Proceedings of the International Conference on computer applications for water supply and distribution*. 8-10 September, 1987. Leicester Polytechnic, UK. Leicester, UK.: Leicester Polytechnic. 1987.

[13] Todini, E. and Pilati, S.. A gradient method for the analysis of pipe networks, in:*Proceedings of the International Conference on computer applications for water supply and distribution.*. 8-10 September, 1987. Leicester Polytechnic, UK.. Leicester, UK.: Leicester Polytechnic. 1987.

[14] Wagner, J.M., Shamier, U. and Marks, D.H. . Water distribution reliability: simulation methods. *Journal of water resources planning and management*, 1988. 114(3): pp. 276-294.